God, Science, and the Big BANG

God, Science, and the Big BANG

a theory of almost everything

James H. Oliver, Jr.

Pleasant Word
A Division of WINEPRESS PUBLISHING

Isaiah 40:31

THEY THAT WAIT UPON
THE LORD

SHALL RENEW THEIR

STRENGTH;
THEY SHALL MOUNT UP
WITH WINGS

AS

EAGLES!

IT IS
HARD
TO
SOAR
WITH
EAGLES

WHEN
YOU'RE
IN A
WORLD
FULL OF
TURKEYS

Printed in the United States of America

Packaged by WinePress Publishing, PO Box 428, Enumclaw, WA 98022. The views expressed or implied in this work do not necessarily reflect those of WinePress Publishing. The author(s) is ultimately responsible for the design, content and editorial accuracy of this work.

ISBN 1-57921-545-9
Library of Congress Catalog Card Number: 2002116740

This book is dedicated to my children; Dan & Kathy Schmeichel, Doug & Kandy Oliver, my special son Jamie, Malcolm & Mary Oliver, Steve & Susan Brammer, Brad & Deb Oliver, Randy & Anne Smith, and all of my grandchildren. They have learned to tolerate an old man who is convinced that believing that the Creator is real is the first step to becoming a Christian and a real scientist.

Special thanks to Anne, Beth, Melody, and Natalie. These four special ladies were instrumental in inviting me to be their Sunday School teacher, and then teaching me more about how to be a Christian than any other teacher or student I have had.

Table Of Contents

Prologue

In the beginning God created the entire Universe. All of it! All matter and energy; but it was smaller than a proton, infinitely hot, and infinitely dense. It was pure force in the beginning. Then God allowed this force to expand. It did not expand like a balloon. It was flat like a discus, that is, thin around the edges and thicker in the middle. As this force expanded it cooled and began to form particles of matter, energy, and light.

At first these particles were packed so close together that they kept bumping into each other and nothing was able to escape from the expanding mass. It was dark and the Spirit was hovering over this expanding mass. The Spirit was watching for just the right moment when all the particles were in the exact place that God wanted them to be. When the mass had expanded to a diameter of about 300,000 light years and cooled to about 3,000 degrees, the Spirit noted that the mass had the density of water and the particles of matter were now in clumps. These clumps were not evenly distributed but were placed through out the expanding mass in just the right position.

Then at precisely the right moment The Word spoke; "**LET THERE BE LIGHT!**" At that exact moment all of the matter and energy of the universe was positioned in the right place so that when The Word spoke, great clumps of matter, energy, and light streamed out from the expanding mass and formed the structures of the Universe.

This may sound like a fairy tale but listen to the words of modern science.

BEHOLD THE HANDWRITING OF GOD

"Astrophysicist and adventurer George Smoot spent twenty years pursuing the 'holy grail of science'—a relentless hunt that led him from the rain forests of Brazil to the frozen wastes of Antarctica. For decades he persevered—struggling against time, the elements, the forces of ignorance and bureaucratic insanity. And finally, on April 23, 1992, he made a startling announcement that would usher in a new scientific age. For George Smoot and his dedicated team of Berkeley researchers had proven the unprovable—uncovering, inarguably and for all time, the secrets of the creation of the Universe."

Purpose

Society in the United States is disintegrating. Society seems to feel that God does not exist or is not relevant to our modern world. A common belief is that life just happened by random chance. Unprovable theories about our origin are taught as "truth".

A column by Brannon S. Howse, in World Net Daily, February 9, 2002, examined the PEERS test put forth by the Nehemiah Institute. This test annually grades the worldview of thousands of adults and school children. According to Howse, "Results from each category are classified into one of four major worldview philosophies: Christian Theism, Moderate Christian, Secular Humanism and Socialism." Howse's findings are staggering. "Christian students attending public schools now regularly score in the lower half of secular humanism, headed toward socialism. Students in typical Christian schools score as secular humanists. Based on projections using the decline rate from Christian students, the church will have lost her posterity to hard-core humanism between 2014–2018."

As a father of seven children and a grandfather of fourteen grandchildren, I feel an immense responsibility to impart a moral and an intellectual heritage to all of my children and grandchildren, as well as all of the rest of the children in our country.

This book explores what we know empirically about the world we live in, and thus what we know about our Creator. We will also compare Biblical statements with what we know from scientific observation. It is my desire that each reader, young or old, be challenged to understand why he or she believes what they believe. As Christians, we must be able to stand firm in our beliefs and respond to challenges of our worldview with an impassioned, inspired, and logical defense.

> 1Peter 3:13–15. "**Who is going to do you wrong if you are devoted to what is good? And yet if you should suffer for your virtues, you may count yourselves happy. Do not fear what they fear: do not be perturbed, but hold the Lord Christ in reverence in your hearts. Be always ready with your defense whenever you are called to account for the hope that is in you, but make that defense with modesty and respect.**"

> 1John 2:28–3:3: "**Even now, my children, dwell in Him, so that when He appears we may be confident and unashamed before Him at His coming. If you know that He is righteous, you must recognize that every man who does right is His child. How great is the love that the Father has shown to us! We are called children of God! Not only called, we really are His children! The reason why the godless world does not recognize us is that it has not known Him. Here and now, dear friends, we are God's children; what we shall be has not yet been disclosed, but we know that when it is disclosed we shall be like Him, because we shall see Him as He is. Everyone who has this hope before him purifies himself, as Christ is pure.**"

Romans 8:14–17: "For all who are moved by the Spirit of God are children of God. The Spirit you have received is not a spirit of slavery leading you back into a life of fear, but a Spirit that makes us children, enabling us to cry 'Abba! Father!' In that cry the Spirit of God joins with our spirit in testifying that we are God's children; and if children, then heirs. We are God's heirs and Christ's fellow heirs, if we share His sufferings now in order to share His splendor hereafter."

My prayer is that this book arms you with knowledge and confidence in the truth of our Lord.

James H. Oliver, Jr. MD

Chapter 1

A Theory Of Almost Everything

Final Reality is that which leaves no further chance for action, discussion, or change. Christianity teaches that Final Reality is a personal, immortal, being whom we call God. He expresses Himself in three persons, the Father, the Son, and The Holy Spirit. Their collective name is "I AM", and their human name is Jesus, The Christ. This God dwells in an area apart from the Universe that is called Eternity or Heaven. It is an area that has no limits and no boundaries. **This God has shaped all matter and energy into its present form by the power of His Word.**

The only alternative view of final reality is that all matter and energy has been shaped into its present form by random chance.

1Corinthians 1:25–31: "Divine folly is wiser than the wisdom of men, and divine weakness is stronger than man's strength. My brothers, think what sort of people you are, whom God has called. Few of you are men of wisdom, by any human standard; few are powerful or highly born. Yet, to shame the wise, God has chosen what the world counts folly, and to shame what is strong, God has chosen what the world counts weakness. He has chosen things low and contemptible, mere nothings, to overthrow the existing order. And so there is no place for human pride in the presence of God. You are in Christ Jesus by God's act, for God has made Him our wisdom; He is our righteousness; in Him we are consecrated and set free. And so in the words of Scripture, "If a Man must boast, let him boast of the Lord."

CHAPTER 2

What Is Faith?

Faith is not just a nebulous feeling. Rather it is the concrete evidence that we are members of God's family. God created humans in His image, but Adam and Eve sinned and humans began the process of dying. Jesus, by His life, death and resurrection opened the way for us to be born over again.

1 John 3: 4. "To commit sin is to break God's law: sin, in fact is lawlessness."

Romans 5:12&15: "Mark what follows. It was through one man that sin entered the world, and through sin death, and thus sin pervaded the whole human race, in as much as all men have sinned." . . . "But God's act of grace is out of all proportion to Adam's wrongdoing. For if the wrongdoing of one man brought death upon so many, its effect is vastly exceeded by the grace of God and the gift that came to so many by the grace of the one man, Jesus Christ."

John 3:5–7: "Jesus answered, 'In truth I tell you, no one can enter the kingdom of God without being born from water and spirit. Flesh can give birth only to flesh; it is spirit that gives birth to spirit. You ought not to be astonished, then, when I tell you that you must be born over again.'"

Recently my daughter-in-law, Deborah Oliver, was complimented by a young man who was a self described Liberal who had been educated at an Ivy League school. He said, "I am amazed at how well read and intelligent you and your family are! For all I knew, you were really unintelligent. After all, you went to some college I'd never even heard of, and you are religious!" (Thank God for Taylor University and other Christian schools and colleges!)

The above compliment demonstrates how the secular world commonly views Christians. It perceives us to be stupid because of our beliefs. Sometimes that view is perpetuated by our own inability to express our belief in, and relationship to, Jesus Christ. At other times a Christian's fear of science, as promoted by the secular world, has made us timid in explaining what science really is. Fear not! The scientific method enables us to learn what the Creator's laws are, and how He works. Real science allows us to learn more about the Creator's power and thus enables us to learn more about Him.

Ignorance has created a deep chasm between human theories and the truth about the origin of our universe. Non-believers try to cross this chasm with a bridge of random chance. Their bridge has no anchor to make it secure. It is full of holes and dangerous. The person who has been "born over again" has a solid foundation-stone that is the anchor for God's bridge across this chasm, and the bridge is the gift of faith from the Creator. Before we delve into science let us examine what the Bible says about God's bridge.

Ephesians 2: 20: "Christ Jesus himself is the foundation-stone."

Galatians 2:17–20: "If now, in seeking to be justified in Christ, we ourselves no less than the gentiles turn out to be sinners against the law, does that mean that Christ is an abettor of sin? No, never! No, if I start building up again a system which I have pulled down, then it is that I show myself up as a transgressor of the law. For through the law I died to law — to live for God. I have been crucified with Christ: the life I now live is not my life, but the life which Christ lives in me; and my present bodily life is lived by the faith of the Son of God, who loved me and gave himself up for me. I will not nullify the grace of God; if righteousness comes by law, then Christ died for nothing."

Hebrews 11:6: "Without faith it is impossible to please Him; for anyone who comes to God must believe that He exists and that He rewards those who search for Him."

Hebrews 11:1–3: "And what is faith? Faith gives substance to our hopes, and makes us certain of realities we do not see. It is for their faith that the men of old stand on record. By faith we perceive that the Universe was fashioned by the Word of God, so that the visible came forth from the invisible."

Romans 1:19&20: "For all that may be known of God by men lies plain before their eyes; indeed God Himself has disclosed it to them. His invisible attributes, that is to say His everlasting power and deity, have been visible, ever since the world began, to the eye of reason, in the things He has made."

Psalm 19:1–6: "The heavens tell out the glory of God, the vault of heaven reveals His handiwork. One day speaks to another, night with night shares its knowledge, and this without speech or language or sound of any voice. Their music goes out through all the earth, their words reach to the ends of the world."

Ephesians 4:10–17: "Finally then, find your strength in the Lord, in His mighty power. Put on all the armor that God provides, so that you may be able to stand firm against the devices of the devil. For our fight is not against human foes, but against cosmic powers, against the authorities and potentates of this dark world, against the superhuman forces of evil in the heavens. Therefore take up God's armor; then you will be able to stand your ground when things are at their worst, to complete every task and still to stand. Stand firm, I say. Fasten on the belt of truth; for coat of mail put on integrity; let the shoes on your feet be the gospel of peace, to give you firm footing; and, with all these, take up the great shield of faith, with which you will be able to quench all the flaming arrows of the evil one. Take salvation for helmet; for sword take that which the spirit gives you, the words that come from God."

It is an absolute certainty that if you have been born over again, you are anchored to the "Foundation-stone". You are a member of God's family! God is your father and Jesus is your brother. Faith is the living, breathing substance of our family ties with God. It is God's gift to us that enables us to understand that He is the Creator of all things. By the faith, which God gives us, we are able to recognize what science really has to say about the Creation, and to see God's power in the Creation.

CHAPTER 3

Communication

At this point some explanations and definitions are needed. Humans are the only creatures on this earth that are able to communicate orally and in writing. However, the words must have the same meaning for the listener or the reader as they do for the speaker or the writer. Otherwise there is no communication, only gibberish.

The following diagram, titled "The logical form of true representations of particular facts", is taken from an article by John W. Oiler Jr. and John L. Omdahl, in an article titled, *Origin of the Human Language Capacity: In Whose Image?* It is contained in the book *The Creation Hypothesis.*[1]

Formal Correspondences

The logical, grammatical, and conceptual elements supplied by intelligence.

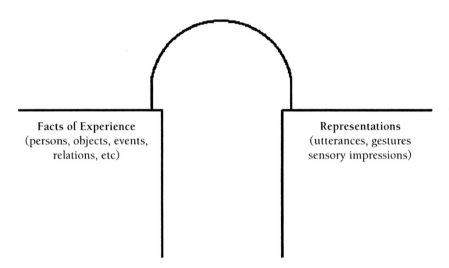

Facts of Experience (persons, objects, events, relations, etc)	**Representations** (utterances, gestures sensory impressions)

The logical form of true representations of particular facts

This diagram shows that a representation that cannot be linked to a sensory experience is a representation that has no meaning. You may say or write that you have etyudf, but if your listener has no sensory experience with etyudf, your representation is meaningless. The same is true of any representation with which your listener has no sensory experience. Just remember what happened at the tower of Babel.

> Genesis 11:1–9: "Now all the world spoke a single language and used few words. As men journeyed in the east, they came upon a plain in the land of Shinar and settled there. They said

to one another, 'Come, let us make bricks and bake them hard'; they used bricks for stone and bitumen for mortar. 'Come,' they said, 'let us build ourselves a city and a tower with its top in the heavens, and make a name for ourselves; or we shall be dispersed all over the earth.' Then the Lord came down to see the city and tower which mortal men had built, and he said, 'Here they are, one people with a single language, and now they have started to do this; henceforward nothing they have a mind to do will be beyond their reach. Come, let us go down there and confuse their speech, so that they will not understand what they say to one another.' So the Lord dispersed them from there all over the earth, and they left off building the city. That is why it is called Babel, because the Lord there made a babble of the language of all the world; from that place the Lord scattered men all over the face of the earth."

Everyone who has knowledge of the findings of modern science should know that modern science has proven that Final Reality (that which is the Ultimate Cause) is something other than impersonal matter or energy shaped into its present form by random chance. That is why most modern scientists avoid any discussion of how the universe and life began. They simply state that the universe has always been or happened by random chance; and that life happened by random chance, or more recently, that it came to earth from somewhere else. They also use complex algebraic equations to change the meaning of terms, and thus make simple things seem complex.

In the following pages of this chapter we will examine various definitions and their validity from a scientific standpoint. These definitions are of extreme importance in interpreting traditional and modern theories about the Universe and life.

Force: Inertia: Matter: Mass: Momentum: Space: Time: Velocity: Weight.

Force is the cause of motion, or change, or stoppage of motion. In the international system of units, the unit of force is the newton. This is the amount of force that causes an object with a mass of 1 kilogram to accelerate 1 meter per second per second. In English units, the unit of force is the poundal, the amount of force that causes an object with a mass of 1 pound to accelerate 1 foot per second per second.

Inertia is the tendency of matter to remain at rest if at rest, or, if moving, to keep moving in the same direction, unless affected by some outside force.

Mass is a quantity of matter forming a body of indefinite size and shape. The mass of a given body is calculated by dividing the weight of the body by its acceleration due to gravity. At sea level, on earth, this is usually expressed as pounds divided by 32 feet per second per second. Mass in physics, is the amount of matter that a given body contains, and one measure of the inertial property of that body, that is, of its resistance to change of motion.

Matter is that which occupies space and can be appreciated by the senses in some fashion.

Momentum is the product of mass times velocity.

Space is the area between particles of matter or quanta of energy.

Time is the period from the beginning to the end of an event.

Velocity is the rate of motion of a given mass. It is measured in terms of distance divided by time, for example, miles per hour.

Weight is a measure of the gravitational attraction for a given mass. Inertial mass and gravitational mass are identical. Weight varies with the position of a given mass relative to a gravitational field; thus, equal masses at the same location in a gravitational field will have equal weights.

One example of a simple concept that has been obscured by changing the meaning of words is; "as velocity approaches the speed of light, mass approaches infinity and time approaches zero." One interpretation of this theorem implies that time will dilate (slow down) and mass will increase as velocity increases. The definitions of the above listed terms will show why this interpretation is incorrect.

Mass in this statement does not change just because its velocity increases. In order for mass to increase, some form of matter must be added to the mass. This would change the amount of matter in the mass, but it would not cause the mass to accelerate. In order for a given mass to change its velocity, force must be applied to the mass. The motion of the mass will then speed up, slow down, change direction, or stop. It depends on the direction in which the force is applied to the mass.

An illustration of force being produced and applied is an explosion. The following example is an eyewitness account of the effect of force. I was sitting in my office at a US Air Force Base Hospital in Kansas. I heard a loud noise that sounded like an explosion in the Base Housing area.

Shortly after the explosion occurred a general alarm was sounded. The emergency response team was directed to the housing area, and I accompanied the ambulance to the site of the explosion. Upon arrival I saw a woman sitting on a sofa in what had been the living room of a house. All of the walls of the house had been pushed outward from the living room. The walls and the furniture were in jumbled piles all around the foundation of the house. The roof was upside down on the lawn, on the side farthest away from the living room. There was no sign of any fire and the woman was not seriously hurt, only frightened.

The investigation revealed that the woman had decided to commit suicide. She had turned off the pilot lights for the gas fired stove and oven in the kitchen of the house, then she turned on all of the gas jets on the stove and in the oven. Afterwards, she went into the living room and waited to die. While she was waiting she decided to smoke one last cigarette. When she attempted to light the cigarette the house exploded. Her only injures were two ruptured ear drums. She had no burns and there was no evidence of any fire in any of the debris left from the explosion. The reason that her injuries were not worse was because the force from an explosion moves outward in all directions from the point of the explosion. The increased air pressure created by the force produced by the explosion caused the ruptured eardrums. Force is not matter in motion; it is what causes matter to move.

An explosion is not the only way to produce force. Any device that produces, changes, or stops motion, must use force to affect the motion. Any practicing engineer knows that for an object to accelerate, decelerate, change direction, or stop moving, it must have some type of force applied to it. Depending on the direction of the application of the force, the object will move forward, backward, change direction, or stop moving. Force may cause motion, but does not add mass to the object being moved.

Any device that can generate force by using some type of fuel will change some of the fuel into force. This will not increase the mass of the accelerating object but will reduce its mass if the object carries its fuel with it. A wind driven object does not lose mass as it accelerates, nor does it gain mass, however it does gain momentum. It is force that causes the air to move producing wind. When the velocity of an object increases, the amount of matter in the object does not increase. In fact the mass of the object will actually decrease if the object uses any of its mass to produce the force that increases the velocity. However the momentum of an object will increase as its velocity increases.

Gravity is an example of negative force (a pull). It is defined as the force that tends to pull all bodies, in the earth's sphere of influence, toward the center of the earth. This pull is a property of all matter and the larger the mass of the pulling body, the greater the force of the pull toward that mass. After seeing that falling objects accelerate toward the earth, Isaac Newton developed the Universal Law of Gravitation. It is routinely used when the space shuttle is launched into orbit around the earth. In words, the law is stated this way: "Every object in the Universe attracts every other object with a force directed along the line of centers for the two objects that is proportional to the product of their masses and inversely proportional to the square of the separation between the two objects." This law can also be expressed in mathematical terms. $F = G \times (m_1 \times m_2 \div r^2)$. F is the gravitational force; m_1 and m_2 are the masses of the two objects; and r is the separation between the objects. G is the universal gravitational constant. It is termed "universal constant" because it is thought to be the same at all places and all times, and thus universally characterizes the intrinsic strength of the gravitational force.

The following article regarding time is taken from the *Microsoft Encarta Encyclopedia2000*. It is important because it help us understand how confused people can become when they try to make time be something that it is not. This article is included to demonstrate that time is the period from the beginning to the end of an event. Clocks are an invention of man and are used to record the duration of segments of time.

"Time; the period during which an action or event occurs; also, a dimension representing a succession of such actions or events. Time is one of the fundamental quantities of the physical world, being similar to length and mass in this respect. Three methods of measuring time are in use at present. The first two methods are based on the daily rotation of the earth on its axis. These methods are determined by the apparent motion of the sun in the sky (solar time) and by the apparent motion of the stars in the sky (sidereal time). The third method of measuring time is based on the revolution of the earth around the sun (ephemeris time)."

"Solar Time; the apparent motion of the sun across the sky has long been used as a basis for measuring time. At any locality, when the sun reaches the highest point in the sky during any given day, it is noon. This point is the meridian. The interval between successive passages of the sun across the same meridian is one day, and this day is by custom divided into 24 hours. The length of the day according to solar time is not the same throughout the year, however, because the apparent motion of the sun varies throughout the year. The difference in the length of the 24-hour day at different seasons of the year can amount to as much as 16 minutes. With the invention of accurate timepieces in the 17th century, this difference became significant. Mean solar time was invented, based on the motion of a hypothetical sun traveling at an even rate throughout the year."

"Standard time, which is based on solar time, was introduced in 1883 by international agreement to avoid the complications that followed in railroad time schedules when each community used its own local solar time. The earth was divided into 24 time zones. The base position is the zero meridian of longitude that passes through the Royal Greenwich Observatory, Greenwich, England. Time zones are described by their distance east or west of Greenwich. Within each time zone all clocks are set to the same time. In the scientific model on which standard time zones are based, each zone spans 15 degrees of longitude; in fact, however, the borders of time zones are bent to conform with state boundaries and international frontiers as well as to facilitate commercial activities. In 1966 the US Congress passed the Uniform Time Act, which established eight standard time zones for the United Sates and its possessions. In 1983 several time zone boundaries were altered so that Alaska, which had formerly spanned four zones, could be nearly unified under one time zone. The US standard time zones are the Atlantic, Eastern, Central, Mountain, Pacific, Alaska, Hawaii-Aleutian, and Samoa zones. In navigation, clocks are often set to the local time at Greenwich, called Greenwich Mean Time

(GMT). Astronomers use this system but call it Universal Time (UT)."

"Sidereal Time: because mean solar time is based on the motion of a hypothetical sun, a base position from which the mean time can be calculated was established. This base position is the vernal equinox, an imaginary point in the sky that is, nevertheless, calculated with great accuracy by astronomers. (The equinox is defined as the point in time when the sun crosses the equator, making night and day of equal duration in all parts of the earth: the vernal equinox occurs about March 21, the autumnal equinox about September 22 or 23.) Practically, the location of the vernal equinox is found by reference to the position of the fixed stars. Solar time based on the position of the stars is called sidereal time. A clock regulated to record sidereal time is called a sidereal clock. The US Naval Observatory in Washington, DC, uses mathematical tables to derive mean solar time from mean sidereal time. The mean solar times thus calculated have a margin of error of 1 part in 1 million. A discrepancy exists between the total number of hours in a mean solar year and in a mean sidereal year because the earth is revolving around the sun at the same time it is rotating on its axis. According to mean sidereal time, the earth returns to the vernal equinox every 365 days 6 hours 9 minutes 9.54 seconds. According to mean solar time, the earth returns every 365 days 5 hours 48 minutes 45.5 seconds; the difference is 20 minutes 24.0 seconds."

"Ephemeris Time: neither mean solar nor mean sidereal time is precisely accurate, because the motion of the earth on its axis is not regular. Variations in the rate of rotation amount to 1 or 2 seconds per year. The rate has varied as much as 30 seconds during the last 200 years. In addition, the earth is gradually slowing down at the rate of about 1/1000 of a second every 100 years. Some of these variations can be taken into account; others cannot because they are irregular. These difficulties were bypassed in 1940 when ephemeris time was introduced. Ephemeris time is used chiefly by astronomers when the greatest degree of accuracy is required in computing the positions of planets and stars. Ephemeris time is based on the annual revolution of the earth around the sun, and the base position, as in sidereal time, is the vernal equinox. Through the use of mathematical tables, ephemeris time is converted into mean solar time."

"The Scientific Standard of Time: until 1955, the scientific standard of time, the second, was based on the earth's period of rotation and was defined as 1/86,400 of the mean solar day. When it was realized that the earth's rate of rotation was irregular and also slowing down, it became necessary to redefine the second. In 1955 the International Astronomical Union defined the second as being 1/31,556,925.9747 of the solar year in progress at noon December 31, 1899. The International Committee on Weights and Measures adopted the definition the following year."

"With the introduction of atomic clocks, specifically, the construction of a high-precision cesium beam atomic clock in 1955, more accurate measurement of time became possible. This atomic clock is tuned to the resonant frequency of the transition energy between two energy states of the cesium-133 atom. In 1967 the measurement of the second in the International System of Units was officially defined as the duration of 9,192,631,770 periods of the radiation corresponding to the transition between two hyperfine levels of the ground state of the cesium-133 atom."

"Time Dilation: time is not a physical constant, although the passage of time in any one place can be measured with great accuracy and precision. The effect of motion and gravity on time is that it is dilated or contracted. In 1905 Albert Einstein formulated the effect of motion on time in his special theory of relativity, and in 1917 he formulated the effect of gravity on time in his general theory of relativity. These effects were observed in experiments conducted in the 1960s and 1970s. In one such experiment in

1971, atomic clocks were carried on two high-speed aircraft (actually they were commercial airliners). One traveled eastward, that is, in the rotational direction of the earth, and one westward. After the flight, the onboard clocks were found to have either lost or gained time (relative to a ground-based atomic clock) depending on their direction of travel, an effect of motion, and their altitude, an effect of gravity. The results confirmed the predictions made in Einstein's theories of relativity."[2]

The above seven articles about time have a few confusing statements in them. The first article is a definition that covers all of time. "<u>Time</u>; the period during which an action or event occurs." This event is the duration of the Universe from the first instant of the creation of the singularity that expanded to become the Universe until it final destruction. The first mistake was to expand the definition to include the statement; "Time is one of the fundamental quantities of the physical world, being similar to length and mass in this respect." First, time is not "one of the fundamental quantities of the physical world, a dimension such as length or width or height". Fundamental quantities are man made absolutes based on measurable standards that can be appreciated by human senses in some manner. Time is not a particle of matter nor is it a form of energy. Time has no length, width, height, or weight. It cannot be measured against any standard. Time is not a thing at all. It is the period from the beginning to the end of an event.

If "time is one of the fundamental quantities of the physical world, being similar to length and mass in this respect", how is it measured? Length, width, and height are measured in meters and fractions or multiples of meters. By international agreement, the standard meter had been defined as the distance between two fine lines on a special bar of platinum-iridium alloy. Length, width, or height is the measurement of something, from one end to the other end. They are measured in comparison to the standard meter. Fundamental quantities of the physical world, that is man made standards, can be measured over and over again to verify the measurement.

The next four articles from the encyclopedia (Solar Time, Standard Time, Sidereal Time, and Ephemeris Time) are man made attempts to standardize time so that it can be measured. They are explanations of different methods of measuring the events called years, days, hours, minutes, and seconds. They are fundamental quantities of the physical world because man-made devices called clocks are used to measure them.

The next article from the encyclopedia (The Scientific Standard of Time) is a definition of one second. The original definition was based on the assumption that one day was 24 hours, one hour was 60 minutes and one minute was 60 seconds. Thus one second was defined as 1/86,400 of the mean solar day. When it was discovered that the rotation of the earth was not regular, and was slowing down, the International Astronomical Union decided to redefine the second. In 1955 the new definition of the second became 1/31,556,925.9747 of the solar year in progress at noon December 31, 1899. With the introduction of a high-precision cesium beam atomic clock in 1955, more accurate recording of time became possible. In 1967 one second in the International System of Units was officially defined as the duration of 9,192,631,770 periods of the radiation corresponding to the transition between two hyperfine levels of the ground state of the cesium-133 atom.

The next article from the encyclopedia (Time Dilation) is an assumption based on Einstein's general and special theories of relativity. In the modern interpretation of these theories the functioning of a clock is confused with time.

This confusion about the functioning of a clock was illustrated in the experiment in which atomic clocks were placed on the top and at the bottom of a water tower. "Another prediction of general relativity is that time should appear to run slower near a massive body like the earth. This prediction was tested in 1962, using a pair of very accurate clocks mounted at the top and bottom of a water tower. The clock at the bottom, which was nearer the earth, was found to run slower, in exact agreement with general relativity.", Notice that the theory of relativity predicted, "That time should appear to run slower near a massive body like the earth." But the actual finding of the experiment was that, "The clock at the bottom, which was nearer the earth, was found to run slower, in exact agreement with general relativity.", Time does not move, but a clock does! Clearly atomic clocks are affected by gravity but time is not.

This confusion about the functioning of a clock and a period of time was also illustrated by this report. "During October 1971, four cesium beam atomic clocks were flown on regularly scheduled commercial jet flights around the world twice, once eastward and once westward to test Einstein's theory of relativity with macroscopic clocks.", The times recorded by the "Flying Clocks" were compared with the (USNO) master clock that was located at the U.S. Naval Observatory. The flying clocks were numbered 120, 361, 408, and 447. The flying clocks lost time (ran slower) on the Eastward flight (120 lost 57 nanoseconds (nsec), 361 lost 74 nsec, 408 lost 55 nsec, and 447 lost 51 nsec.) These same clocks gained time (ran faster) on the westward flight, (120 gained 277 nsec, 361 gained 284 nsec, 408 gained 266 nsec, and 447 gained 266 nsec). Obviously the functioning of the flying clocks was affected by gravity, motion, altitude, and or the direction of the flight. Time did not slow down or speed up during the flights but the functioning of the clocks certainly did.

This functioning of the atomic clocks is well known by the controllers who operate the Global Positioning System. The clocks in the satellites that are used by this system are monitored twice daily, and the clocks are routinely adjusted to match the time on the master clock at Falcon Air Force Base, Colorado. This adjustment is necessary because the satellite clocks are not in the same gravitational field as the clock at Falcon Air Force Base.

The proper way to understand the relationship between velocity, mass and time is easily illustrated. (The amount of momentum on arrival in this illustration is approximate. To be very accurate, the mass of the fuel used up would have to be deducted from the original mass and the remainder then multiplied by the velocity.)

It is fifteen miles from my house to my office. If my car and I have a mass of 1500 pounds, and if I drive from my house to my office at a steady fifteen miles per hour (15 miles per 3600 seconds), it would take one hour to get there, and my momentum (mass x velocity) on arrival would be 22,500 pounds. It would take a force of 22,500 poundals to stop the motion of the car.

If I drove at four times that velocity (sixty miles per 3600 seconds), it would take fifteen minutes to get there, and my momentum on arrival would be 90,000 pounds.

If I drove at forty times that velocity (six hundred miles per 3600 seconds), it would take 90 seconds to get there, and my momentum on arrival would be 900,000 pounds; and if I drove at 400 times that velocity (six thousand miles per 3600 seconds), it would take 9 seconds to get there, and my momentum on arrival would be 9,000,000 pounds.

Simply stated the relationship between velocity, momentum, and time is: "the faster you go the less time it takes to get where you are going, and the harder it is to stop when you get there." If you could go at infinite velocity, it would take no time at all to get where you were going, and it would take an infinite force to stop you. The relationship between time, space, and velocity will become very important when we discuss how much time it took for the Universe to form.

One more fact to remember the speed of light is estimated to be about 186,000 miles per second. A light year is a measure of distance, in miles, traveled at a velocity of 186,000 miles per second for 365 days, or (186,000 times 365 times 24 times 60 times 60) miles. This is 5,865,696,000,000 miles, usually rounded off to

5.88 times 10^{12} miles. A light year is not a measure of time, it is a measure of distance based on the assumption that light always travels through outer space at 186,000 miles per second, and that one second is always the same.

Remember, before Creation, the Universe did not exist. Because there had not yet been a beginning, time had not begun. No matter existed. When the Universe began to expand, it did not expand to fill empty space. Space is simply the empty area between the particles of matter that make up the Universe. Space is not a thing; rather it is the absence of anything between these particles of matter. Time began the moment the Universe began to expand, and will continue as long as the Universe exists. The most recent scientific evidence indicates that the original expansion (commonly referred to as the Big Bang) occurred faster than the speed of light. If this is true, then the formation of the universe may have taken almost no time at all!

How do we know how old the universe is? Actually we have no empirical scientific evidence to give us any idea about the age of the earth. These data do not exist because no human was on hand to observe and record the events. What does the Bible have to say about time, or is time simply not important to knowing anything about God? It seems that God wants us to know that time had a beginning but does not want us to know when time will end.

Genesis 1:1: "**In the beginning God created heaven and Earth**".

2 Peter 3:8: "**With the Lord one day is like a thousand years and a thousand years like one day.**"

Matthew 24:26: "**But about that day and hour no one knows, not even the angels in heaven, not even the Son; only the Father.**"

Since no known culture counts days from evening to morning, the expression in Genesis chapter 1, "evening came and morning came, the___day", is best understood to indicate a period of time. The one basic thing to remember; time is not a tangible thing. Rather it is a continuum, and can be defined as the period from the Creation to the final destruction of the Universe. Time on earth is recorded, based on two events: the period during which the earth revolves once on its axis (high noon to the next high noon) and the period during which the earth revolves once around the sun. One revolution of the earth on its axis is called one day. One revolution of the earth around the sun is recorded in days or months with one complete trip around the sun being called one year. We chop the time continuum into segments (hours, minutes, and seconds) so that we can measure the duration of an event as it happens in our world.

If we knew when the fusion reaction started in the sun; how much hydrogen was present when the fusion reaction started; how much hydrogen is being fused each second; if this amount is constant; and if a second was constant, then we could predict when the sun would expand into a "red giant" and then collapse into a white dwarf. Life on earth and the earth itself would end. Fortunately, we do not know the answer to any of these questions, but Jesus knew what He was talking about when He stated:

Matthew 24:36: **"But about that day and hour no one knows, not even the angels in heaven, not even the Son; only the Father."**

In order to understand the Biblical use of the terms light and darkness, life and death, good and evil, why time is important, and how they are all related we need to understand these verses.

1 John 1:5: "Here is the message we heard from Him (Jesus) and pass on to you: that God is Light, and in Him there is no darkness at all."

John 1:4&5: "All that came to be was alive with His life, and that life was the light of men. The light shines on in the dark, and the darkness has never mastered it."

John 3:19–21: "Here lies the test; the light has come into the world, but men preferred darkness to light because their deeds were evil. Bad men all hate the light and avoid it, for fear their practices should be shown up. The honest man comes to the light so that it may be clearly seen that God is in all that he does."

Ezekiel 18: 4 & 20: "The soul that sins shall die."

Then from the writings of St. Augustine, if God created everything, then evil is not something. It is the absence of something that should be there. Evil is the absence of the Spirit of God just as darkness is the absence of light, death is the absence of life, and space is the absence of matter. It is always Light where God's Spirit is. Darkness is where God's Spirit is not! In the same way if you have been "born over again" then God's Spirit is living in you and you are alive because the Spirit of God is living in you. If you have not been "born over again" you are still dwelling in darkness in spite of the "Light" that has come into the world.

John 3:3: "In truth, in very truth I tell you, unless a man has been born over again he cannot see the kingdom of God."

When you are in total darkness, you really cannot see anything! If you want to be able to see and to live forever, you must be born over again!!

When you are in total darkness, you really cannot see anything!
If you want to be able to see and to live forever,
you must be born over again!!

CHAPTER 4

Time Illustrated

Genesis 1:1: "In the beginning God created Heaven and earth"

2 Peter 3:10: "On that day the heavens will disappear with a great rushing sound, the elements will disintegrate in flames, and the earth and all that is in it will be burnt up."

This drawing represents time from the Creation to the final destruction of the Universe.

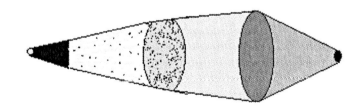

The white dot represents Creation, the singularity from which the universe grew. The black cone represents Genesis 1:2: the period of time from the beginning until the photons were able to escape from the initial "soup of matter". The small dots in the white cone represent Genesis 1:3-5: the escaping photons and the background radiation, left over from the initial expansion. The white oval with the speckled clusters represents the universe as it is now. The speckles are the stars and galaxies placed as God determined in the beginning. The white, clear area is the future and the black oval represents 2 Peter 3:10: "The Day of The Lord" when the entire universe will be "burned up". The black dot represents Revelation 21:1: "the first heaven and the first earth had vanished". It is the final "black hole" which is the end of this present universe.

On the following pages of this book we will show that it is smart to believe that the Creator really exists. From the creation, we will learn that He is powerful beyond our wildest imagination. We will also learn that the God of the Bible is the Creator, and if we want to know how God works with humans we need to study the Bible and how it points us to Jesus.

CHAPTER 5

GOD The Creator

If there is a God who is the CREATOR of all things and who shaped all matter and energy into its present form, then He must have existed prior to creation of this universe. Since the Creator must have existed prior to creating this universe, He cannot be part of this universe. Since humans are part of this creation, and God is other than this creation, the only way we can know anything about Him is for this Creator God to reveal Himself to us. Christianity teaches that there is such a God, the Creator, whose name is I AM. It also teaches that He has revealed Himself to us.

Logic, the science which deals with the criteria of valid thought, requires that this revelation must be true, and in a form that we can understand. If the form of the revelation is not understandable then it is not revelation. It is just random "noise". If the form of the revelation is understandable but not true, then God does not want us to know anything about Himself, or He is simply playing games with us.

Christians believe that the revelation that God has given us is both understandable and true. It tells us whom God is, how He operates, and what He is like, and the record of Creation has been written by God himself. Examples of "The Hand Writing of God" are the background radiation left over from the first moments of creation, the complex structure of the universe, and the chemical, deoxyribonucleic acid (DNA).

Christians also believe that God never lies. When human theories conform to the revelation that God has given us, the theories are true. When human theories do not conform to the revelation that God has given to us, the theories are not true. Darwinian Evolution, as a worldview, is an example of a false theory. It dictates that all the different kinds of living animals gradually developed from the very simple to the more complex; that this development occurred by random chance; and that it took several million years to happen.

The fossils at the Chengjiang site in China are all in the same geologic layer. (This scientific evidence is clearly a revelation from the Creator.) In this geologic layer there are fossils of every phylum of animal that exists on earth today, as well as at least twelve phyla of animals that no longer exist. The theory of Evolution does not conform to this revelation. The fossil evidence demonstrates that development of life forms, from very simple to very complex did not occur over millions of years. All of the life forms appeared in the same geologic era. This evidence is called the Cambrian Explosion of life. It has been documented by many scientists, one of who is Dr. Paul Chien, chairman of the Biology Department at the University of San Francisco.

In *The Real Issue* Dr. Chien had this to say about the fossils at the Chengjiang site in China and similar fossils at the Burgess Shale site in Canada. "A simple way of putting it is that currently we have about 38 phyla of different groups of animals, but the total number of phyla discovered during that period of time (the Cambrian Explosion of Life) including those in China, Canada, and elsewhere, adds up to over 50 phyla. That means there more phyla in the very, very beginning, where we found the first fossils of animal life, than exist now . . .

The general impression people get is that we began with micro-organisms, then came lowly animals that don't amount to much, and then came the birds, mammals and man. Scientists were looking at a very small branch of the whole animal kingdom, and they saw more complexity and advanced features in that group. But it turns out that this concept does not apply to the entire spectrum of animals or to the appearance or creation of different groups. Take all the different body plans of roundworms, flatworms, coral, jellyfish and whatever—all those appeared at the very first instant . . . Since the Cambrian period, we have only die-off and no new groups coming about, ever."[5] In addition no intermediate forms were found in any of these sites.

When the revelation that God has given to humans is silent about human theories, these theories are unimportant for knowing about God.

Christians believe that God has revealed Himself to us in at least three ways.

1. God has revealed Himself through what He has created.

Romans 1:19&20: "For all that may be known of God by men lies plain before their eyes; indeed God Himself has disclosed it to them. His invisible attributes, that is to say His everlasting power and deity, have been visible, ever since the world began, to the eye of reason, in the things He has made."

Psalm 19:1–4: "The heavens tell out the glory of God, the vault of heaven reveals His handiwork. One day speaks to another, night with night shares its knowledge, and this without speech or language or sound of any voice. Their music goes out through all the earth, their words reach to the end of the world."

If the creation shows God's "**everlasting power and deity**", and this has "**been visible ever since the world began, to the eye of reason, in the things He has made**" and "**One day speaks to another, night with night shares its knowledge, and this without speech or language or sound of any voice**", we should be able read this "Hand Writing of God" in His Creation. In the twentieth century we were finally beginning to learn how to do this.

2. God has also revealed Himself in the form of a man to show humans what He is like.

John 1:1–3:"**When all things began, the Word already was. The Word dwelt with God, and what God was, the Word was. The Word, then, was with God at the beginning, and through Him all things came to be; no single thing was created without Him. All that came to be alive was alive with his life, and that life was the light of men.**"

Hebrews 13:8: "**Jesus Christ is the same yesterday, today, and forever**".

Christians believe that <u>Jesus Christ is God, the Creator of all things, and He never changes.</u> Since He made all things, the "laws of nature" are His laws, and all of creation obeys these laws! This includes the activity involved in creating as well as the creation itself. It does not mean that we know all of these laws (but God knows them since He made them), only that they are knowable if we use the "eye of reason" to discover them. These laws are reasonable and knowable, and God has given to humans the ability to know these laws <u>if humans believe Him. If humans do not know Him or do not believe Him they will never know the true "laws of nature" as created by God who is The Creator.</u>

3. God has also given to humans a true record, written by humans, of what He has done in and with His creation.

2Timothy 3:16&17: **"Every inspired scripture has its use for teaching the truth and refuting error, or for reformation of manners and discipline in right living so that the man who belongs to God may be efficient and equipped for good works of every kind."** The proof that this record is true is contained in the chapter titled Admissible Evidence.

Finally, God is actively involved on a daily basis with His children. This involvement is clear evidence that He is not an "absentee father"; He cares about every aspect of our existence.

Franky Schaeffer V (Francis A Schaeffer's son) in *Addicted to Mediocrity* summarizes who and what God is and is not. "Either God is the Creator of the whole man, the whole universe, and all of reality and existence, or He is the Creator of none of it. If God is only the creator of some divided platonic existence, which leads to a tension between the body and soul, the real world and the spiritual world, if God is only the Creator of some spiritual little experiential 'praise the Lord' reality, then He is not much of a God. In deed He is not I AM at all. If our Christian lives are allowed to become something spiritual and religious as opposed to something real, daily applicable, understandable, beautiful, verifiable, balanced, sensible, and above all united, whole; if indeed our Christianity is allowed to become this waffling spiritual goo that nineteenth—century platonic Christianity became, then Christianity as truth disappears and instead we only have a system of vague experiential religious platitudes in its place."[6]

CHAPTER 6

Random Chance

We have talked about God, the Creator, being Final Reality, but there is another worldview that is very prevalent in our schools and Universities. The primary statement of this other worldview is that all matter and energy has been shaped into its present form by random chance. There are several variations of this theory, but first the general idea itself.

The first theory about random chance and the Universe was best stated by Carl Sagan at the opening of the television program *Cosmos*. "The Cosmos is all that is, all that ever has been, and all that ever will be." This is a statement of faith and cannot be discussed. To believe this means that there is no beginning, no end, and no purpose to anything. Ultimately it come down to this, "Eat, drink, and be merry for tomorrow we die." Nothing matters. There are no absolutes, no right or wrong, and anything is OK since everything is only random chance. Many people in this country have accepted this worldview, which explains why our society is in such a state of decay.

One variation of this theory is called theistic evolution, a classic oxymoron. An oxymoron is a figure of speech in which opposite or contradictory ideas or terms are combined. Theism is defined in Webster's Unabridged Dictionary as: "belief in one God who is creator and ruler of the universe and known by revelation". Evolution is atheistic and teaches, "The universe is self-existing and not created." Those who hold the theistic evolution worldview believe that something was the creator, but what ever it was, it used random chance as its way of creating. I believe that the following verses apply to this worldview.

Revelation 3:14–22: "To the angel of the church at Laodicea write: 'These are the words of the Amen, the faithful and true witness, the prime source of all of God's creation: I know all your ways; you are neither hot nor cold. How I wish you were either hot or cold! But because you are lukewarm, neither hot nor cold, I will spit you out of my mouth. You say, how rich I am! And how well I have done! I have everything I want. In fact, thought you do not know it. You are the most pitiful wretch, poor, blind, and naked. So I advise you to buy from Me gold refined in the fire, to make you truly rich; and white clothes to put on to hide the shame of your nakedness; and ointment for your eyes that you may see. All whom I love I reprove and discipline. Be on your mettle therefore and repent. Here I stand knocking at the door; if anyone hears my voice and opens the door, I will come in and sit down to supper with him and he with me. To him who is victorious I will grant a place on my throne, as I myself was victorious and sat down with my Father on His throne. Hear, you who have ears to hear, what the spirit says to the churches."

With the exception of a few Christian schools, Christian colleges, and some home schools, the vast majority of public and private schools and colleges in the United States are now teaching a different variation of the random chance worldview. It is identified by a variety of different terms: pragmatism, Secular Humanism, materialism, atheistic evolution, neodarwinism, nihilism, or methodological naturalism. To avoid confusion, we will use the term Secular Humanism as being synonymous with all the other terms because all of these terms apply to the same worldview.

In 1961 the United States Supreme Court, in *Torcase v. Watkins,* specifically defined Secular Humanism as a religion equivalent to theistic and other non-theistic religions. Then, with total disregard for the First Amendment to the Constitution, the Supreme Court and most of the federal courts have endorsed the teaching of Secular Humanism in the public schools, and declared it completely legal, while at the same time they have outlawed the teaching of Christianity and some other religions. The worldview of Secular Humanism is; final reality is impersonal matter or energy shaped into its present form by random chance; and human ideas as expressed by Humanists are the only source of truth. The first statement of Humanist doctrine, *The Humanist Manifesto I* was published in 1933. The revised edition, *Humanist Manifesto II* was published in 1973. The most recent edition is called *A Secular Humanist Declaration* and was published in 1980. Its most recent update was on 11/10/2000.

I have included a copy of *A Secular Humanist Declaration* because it is the most commonly taught worldview in our public schools and Universities.

A Secular Humanist Declaration

Secular humanism is a vital force in the contemporary world. It is now under unwarranted and intemperate attack from various quarters. This declaration defends only that form of secular humanism, which is explicitly committed to democracy. It is opposed to all varieties of belief that seek supernatural sanction for their values or espouse rule by dictatorship.

Democratic secular humanism has been a powerful force in world culture. Its ideals can be traced to the philosophers, scientists, and poets of classical Greece and Rome, to ancient Chinese Confucian society, to the Carvaka movement of India, and to other distinguished intellectual and moral traditions. Secularism and humanism were eclipsed in Europe during the Dark Ages, when religious piety eroded humankind's confidence in its own powers to solve human problems. They reappeared in force during the Renaissance with the reassertion of secular and humanist values in literature and the arts, again in the sixteenth and seventeenth centuries with the development of modern science and a naturalistic view of the universe, and their influence can be found in the eighteenth century in the Age of Reason and the Enlightenment.

Democratic secular humanism has creatively flowered in modern times with the growth of freedom and democracy. Countless millions of thoughtful persons have espoused secular humanist ideals, have lived significant lives, and have contributed to the building of a more humane and democratic world. The modern secular humanist outlook has led to the application of science and technology to the improvement of the human condition. This has had a positive effect on reducing poverty, suffering, and disease in various parts of the world, in extending longevity, on improving transportation and communication, and in making the good life possible for more and more people. It has led to the emancipation of hundreds of millions of people from the exercise of blind faith and fears of superstition and has contributed to their education and the enrichment of their lives.

Secular humanism has provided an impetus for humans to solve their problems with intelligence and perseverance, to conquer geographic and social frontiers, and to extend the range of human exploration and adventure. Regrettably, we are today faced with a variety of antisecularist trends: the reappearance of dogmatic authoritarian religions; fundamentalist, literalist, and doctrinaire Christianity; a rapidly growing and uncompromising Moslem clericalism in the Middle East and Asia; the reassertion of orthodox authority by the Roman Catholic papal hierarchy; nationalistic religious Judaism; and the reversion to obscurantist religions in Asia.

New cults of unreason as well as bizarre paranormal and occult beliefs, such as belief in astrology, reincarnation, and the mysterious power of alleged psychics, are growing in many Western societies. These disturbing developments follow in the wake of the emergence in the earlier part of the twentieth century of intolerant messianic and totalitarian quasi-religious movements, such as fascism and communism. These religious activists not only are responsible for much of the terror and violence in the world today but also stand in the way of solutions to the world's most serious problems.

Paradoxically, some of the critics of secular humanism maintain that it is a dangerous philosophy. Some assert that it is "morally corrupting" because it is committed to individual freedom, others that it condones "injustice" because it defends democratic due process. We who support democratic secular humanism deny such charges, which are based upon misunderstanding and misinterpretation, and we seek to outline a set of principles that most of us share.

Secular humanism is not a dogma or a creed. There are wide differences of opinion among secular humanists on many issues. Nevertheless, there is a loose consensus with respect to several propositions. We are apprehensive that modern civilization is threatened by forces antithetical to reason, democracy, and freedom. Many religious believers will no doubt share with us a belief in many secular humanist and democratic values, and we welcome their joining with us in the defense of these ideals.

1. *Free Inquiry*

The first principle of democratic secular humanism is its commitment to free inquiry. We oppose any tyranny over the mind of man, any efforts by ecclesiastical, political, ideological, or social institutions to shackle free thought. In the past, such tyrannies have been directed by churches and states attempting to enforce the edicts of religious bigots. In the long struggle in the history of ideas, established institutions, both public and private, have attempted to censor inquiry, to impose orthodoxy on beliefs and values, and to excommunicate heretics and extirpate unbelievers. Today, the struggle for free inquiry has assumed new forms. Sectarian ideologies have become the new theologies that use political parties and governments in their mission to crush dissident opinion. Free inquiry entails recognition of civil liberties as integral to its pursuit, that is, a free press, freedom of communication, the right to organize opposition parties and to join voluntary associations, and freedom to cultivate and publish the fruits of scientific, philosophical, artistic, literary, moral and religious freedom. Free inquiry requires that we tolerate diversity of opinion and that we respect the right of individuals to express their beliefs, however unpopular they may be, without social or legal prohibition or fear of sanctions. Though we may tolerate contrasting points of view, this does not mean that they are immune to critical scrutiny. The guiding premise of those who believe in free inquiry is that truth is more likely to be discovered if the opportunity exists for the free exchange of opposing opinions; the process of interchange is frequently as important as the result. This applies not only to science and to everyday life, but also to politics, economics, morality, and religion.

2. *Separation Of Church And State*

Because of their commitment to freedom, secular humanists believe in the principle of the separation of church and state. The lessons of history are clear: wherever one religion or ideology is established and given a dominant position in the state, minority opinions are in jeopardy. A pluralistic, open democratic society allows all points of view to be heard. Any effort to impose an exclusive conception of Truth, Piety, Virtue, or Justice upon the whole of society is a violation of free inquiry. Clerical authorities should not be permitted to legislate their own parochial views — whether moral, philosophical, political, educational, or social — for the rest of society. Nor should tax revenues be exacted for the benefit or support of sectarian religious institutions. Individuals and voluntary associations should be free to accept or not to accept any belief and to support these convictions with whatever resources they may have, without being compelled by taxation to contribute to those religious faiths with which they do not agree. Similarly, church properties should share in the burden of public revenues and should not be exempt from taxation. Compulsory religious oaths and prayers in public institutions (political or educational) are also a violation of the separation principle. Today, nontheistic as well as theistic religions compete for attention. Regrettably, in communist countries, the power of the state is being used to impose an ideological doctrine on the society, without tolerating the expression of dissenting or heretical views. Here we see a modern secular version of the violation of the separation principle.

3. The Ideal Of Freedom

There are many forms of totalitarianism in the modern world — secular and nonsecular — all of which we vigorously oppose. As democratic secularists, we consistently defend the ideal of freedom, not only freedom of conscience and belief from those ecclesiastical, political, and economic interests that seek to repress them, but genuine political liberty, democratic decision making based upon majority rule, and respect for minority rights and the rule of law. We stand not only for freedom from religious control but for freedom from jingoistic government control as well. We are for the defense of basic human rights, including the right to protect life, liberty, and the pursuit of happiness. In our view, a free society should also encourage some measure of economic freedom, subject only to such restrictions as are necessary in the public interest. This means that individuals and groups should be able to compete in the marketplace, organize free trade unions, and carry on their occupations and careers without undue interference by centralized political control. The right to private property is a human right without which other rights are nugatory. Where it is necessary to limit any of these rights in a democracy, the limitation should be justified in terms of its consequences in strengthening the entire structure of human rights.

4. *Ethics Based On Critical Intelligence*

The moral views of secular humanism have been subjected to criticism by religious fundamentalist theists. The secular humanist recognizes the central role of morality in human life; indeed, ethics was developed as a branch of human knowledge long before religionists proclaimed their moral systems based upon divine authority. The field of ethics has had a distinguished list of thinkers contributing to its development: from Socrates, Democritus, Aristotle, Epicurus, and Epictetus, to Spinoza, Erasmus, Hume, Voltaire, Kant, Bentham, Mill, G. E. Moore, Bertrand Russell, John Dewey, and others. There is an influential philosophical tradition that maintains that ethics is an autonomous field of inquiry that ethical judgments can be formulated independently of revealed religion, and that human beings can cultivate practical reason and wisdom and, by its application, achieve lives of virtue and excellence. Moreover, philosophers have emphasized the need to cultivate an appreciation for the requirements of social justice and for an individual's obligations and responsibilities toward others. Thus, secularists deny that morality needs to be deduced from religious belief or that those who do not espouse a religious doctrine are immoral. For secular humanists, ethical conduct is, or should be, judged by critical reason, and their goal is to develop autonomous and responsible individuals, capable of making their own choices in life based upon an understanding of human behavior. Morality that is not God based needs not be antisocial, subjective, or promiscuous, nor need it lead to the breakdown of moral standards. Although we believe in tolerating diverse lifestyles and social manners, we do not think they are immune to criticism. Nor do we believe that any one church should impose its views of moral virtue and sin, sexual conduct, marriage, divorce, birth control, or abortion, or legislate them for the rest of society. As secular humanists we believe in the central importance of the value of human happiness here and now. We are opposed to absolutist morality, yet we maintain that objective standards emerge, and ethical values and principles may be discovered, in the course of ethical deliberation. Secular humanist ethics maintains that it is possible for human beings to lead meaningful and wholesome lives for themselves and in service to their fellow human beings without the need of religious commandments or the benefit of clergy. There have been any number of distinguished secularists and humanists who have demonstrated moral principles in their personal lives and works: Protagoras, Lucretius, Epicurus, Spinoza, Hume, Thomas Paine, Diderot, Mark Twain, George Eliot, John Stuart Mill, Ernest Renan, Charles Darwin, Thomas Edison, Clarence Darrow, Robert Ingersoll, Gilbert Murray, Albert Schweitzer, Albert Einstein, Max Born, Margaret Sanger, and Bertrand Russell, among others.

5. *Moral Education*

We believe that moral development should be cultivated in children and young adults. We do not believe that any particular sect can claim important values as their exclusive property; hence it is the duty of public education to deal with these values. Accordingly, we support moral education in the schools that is designed to develop an appreciation for moral virtues, intelligence, and the building of character. We wish to encourage wherever possible the growth of moral awareness and the capacity for free choice and an understanding of the consequences thereof. We do not think it is moral to baptize infants, to confirm adolescents, or to impose a religious creed on young people before they are able to consent. Although children should learn about the history of religious moral practices, these young minds should not be indoctrinated in a faith before they are mature enough to evaluate the merits for themselves. It should be noted that secular humanism is not so much a specific morality as it is a method for the explanation and discovery of rational moral principles.

6. Religious Skepticism

As secular humanists, we are generally skeptical about supernatural claims. We recognize the importance of religious experience: that experience that redirects and gives meaning to the lives of human beings. We deny, however, that such experiences have anything to do with the supernatural. We are doubtful of traditional views of God and divinity. Symbolic and mythological interpretations of religion often serve as rationalizations for a sophisticated minority, leaving the bulk of mankind to flounder in theological confusion. We consider the universe to be a dynamic scene of natural forces that are most effectively understood by scientific inquiry. We are always open to the discovery of new possibilities and phenomena in nature. However, we find that traditional views of the existence of God are meaningless, have not yet been demonstrated to be true, or are tyrannically exploitative. Secular humanists may be agnostics, atheists, rationalists, or skeptics, but they find insufficient evidence for the claim that some divine purpose exists for the universe. They reject the idea that God has intervened miraculously in history or revealed himself to a chosen few or that he can save or redeem sinners. They believe that men and women are free and are responsible for their own destinies and that they cannot look toward some transcendent Being for salvation. We reject the divinity of Jesus, the divine mission of Moses, Mohammed, and other latter day prophets and saints of the various sects and denominations. We do not accept as true the literal interpretation of the Old and New Testaments, the Koran, or other allegedly sacred religious documents, however important they may be as literature. Religions are pervasive sociological phenomena, and religious myths have long persisted in human history. In spite of the fact that human beings have found religions to be uplifting and a source of solace, we do not find their theological claims to be true. Religions have made negative as well as positive contributions toward the development of human civilization. Although they have helped to build hospitals and schools and, at their best, have encouraged the spirit of love and charity, many have also caused human suffering by being intolerant of those who did not accept their dogmas or creeds. Some religions have been fanatical and repressive, narrowing human hopes, limiting aspirations, and precipitating religious wars and violence. While religions have no doubt offered comfort to the bereaved and dying by holding forth the promise of an immortal life, they have also aroused morbid fear and dread. We have found no convincing evidence that there is a separable "soul" or that it exists before birth or survives death. We must therefore conclude that the ethical life can be lived without the illusions of immortality or reincarnation. Human beings can develop the self-confidence necessary to ameliorate the human condition and to lead meaningful, productive lives.

7. *Reason*

We view with concern the current attack by nonsecularists on reason and science. We are committed to the use of the rational methods of inquiry, logic, and evidence in developing knowledge and testing claims to truth. Since human beings are prone to err, we are open to the modification of all principles, including those governing inquiry, believing that they may be in need of constant correction. Although not so naive as to believe that reason and science can easily solve all human problems, we nonetheless contend that they can make a major contribution to human knowledge and can be of benefit to humankind. We know of no better substitute for the cultivation of human intelligence.

8. *Science And Technology*

We believe the scientific method, though imperfect, is still the most reliable way of understanding the world. Hence, we look to the natural, biological, social, and behavioral sciences for knowledge of the universe and man's place within it. Modern astronomy and physics have opened up exciting new dimensions of the universe: they have enabled humankind to explore the universe by means of space travel. Biology and the social and behavioral sciences have expanded our understanding of human behavior. We are thus opposed in principle to any efforts to censor or limit scientific research without an overriding reason to do so. While we are aware of, and oppose, the abuses of misapplied technology and its possible harmful consequences for the natural ecology of the human environment, we urge resistance to unthinking efforts to limit technological or scientific advances. We appreciate the great benefits that science and technology (especially basic and applied research) can bring to humankind, but we also recognize the need to balance scientific and technological advances with cultural explorations in art, music, and literature.

9. *Evolution*

Today the theory of evolution is again under heavy attack by religious fundamentalists. Although the theory of evolution cannot be said to have reached its final formulation, or to be an infallible principle of science, it is nonetheless supported impressively by the findings of many sciences. There may be some significant differences among scientists concerning the mechanics of evolution; yet the evolution of the species is supported so strongly by the weight of evidence that it is difficult to reject it. Accordingly, we deplore the efforts by fundamentalists (especially in the United States) to invade the science classrooms, requiring that creationist theory be taught to students and requiring that it be included in biology textbooks. This is a serious threat both to academic freedom and to the integrity of the educational process. We believe that creationists surely should have the freedom to express their viewpoint in society. Moreover, we do not deny the value of examining theories of creation in educational courses on religion and the history of ideas; but it is a sham to mask an article of religious faith as a scientific truth and to inflict that doctrine on the scientific curriculum. If successful, creationists may seriously undermine the credibility of science itself.

10. *Education*

In our view, education should be the essential method of building humane, free, and democratic societies. The aims of education are many: the transmission of knowledge; training for occupations, careers, and democratic citizenship; and the encouragement of moral growth. Among its vital purposes should also be an attempt to develop the capacity for critical intelligence in both the individual and the community. Unfortunately, the schools are today being increasingly replaced by the mass media as the primary institutions of public information and education. Although the electronic media provide unparalleled opportunities for extending cultural enrichment and enjoyment, and powerful learning opportunities, there has been a serious misdirection of their purposes. In totalitarian societies, the media serve as the vehicle of propaganda and indoctrination. In democratic societies television, radio, films, and mass publishing too often cater to the lowest common denominator and have become banal wastelands. There is a pressing need to elevate standards of taste and appreciation. Of special concern to secularists is the fact that the media (particularly in the United States) are inordinately dominated by a pro religious bias. The views of preachers, faith healers, and religious hucksters go largely unchallenged, and the secular outlook is not given an opportunity for a fair hearing. We believe that television directors and producers have an obligation to redress the balance and revise their programming. Indeed, there is a broader task that all those who believe in democratic secular humanist values will recognize, namely, the need to embark upon a long-term program of public education and enlightenment concerning the relevance of the secular outlook to the human condition.

Conclusion

Democratic secular humanism is too important for human civilization to abandon. Reasonable persons will surely recognize its profound contributions to human welfare. We are nevertheless surrounded by doomsday prophets of disaster, always wishing to turn the clock back; they are anti science, anti freedom, anti human. In contrast, the secular humanistic outlook is basically melioristic, looking forward with hope rather than backward with despair. We are committed to extending the ideals of reason, freedom, individual and collective opportunity, and democracy throughout the world community. The problems that humankind will face in the future, as in the past, will no doubt be complex and difficult. However, if it is to prevail, it can only do so by enlisting resourcefulness and courage. Secular humanism places trust in human intelligence rather than in divine guidance. Skeptical of theories of redemption, damnation, and reincarnation, secular humanists attempt to approach the human situation in realistic terms: human beings are responsible for their own destinies. We believe that it is possible to bring about a more humane world, one based upon the methods of reason and the principles of tolerance, compromise, and the negotiations of difference.

We recognize the need for intellectual modesty and the willingness to revise beliefs in the light of criticism. Thus consensus is sometimes attainable. While emotions are important, we need not resort to the panaceas of salvation, to escape through illusion, or to some desperate leap toward passion and violence. We deplore the growth of intolerant sectarian creeds that foster hatred. In a world engulfed by obscurantism and irrationalism it is vital that the ideals of the secular city not be lost.

A Secular Humanist Declaration was drafted by Paul Kurtz, Editor of *Free Inquiry*.

Humanism is a religion that clearly teaches that God does not exist, and Jesus warned us about these types of teachers.

Matthew 15:14: "**Leave them alone; they are blind guides of blind men, and if one blind man guides another they will both fall into the ditch.**"

In any battle, if you want to win, you must know your enemy. Secular Humanism clearly is the enemy of Biblical and scientific truth!

CHAPTER 7

Mathematics

There is another worldview based on the belief that mathematical equations are the only trustworthy source of truth. The majority of theoretical physicists and science teachers have accepted this worldview, but it results in a strange mixture of illusions and two-dimensional frames of reference. This world view is like the opening statement of chapter 2 of *Alice's Adventures in Wonderland*: "Curiouser and curiouser!"

Charles Darwin wrote a book, *The Origin of the Species*, that was the pivotal changing point for the way the western world thought about the origin of the Universe and the origin of life. Later another young man who was extremely clever at manipulating mathematical equations, proposed a new theory of reality based on the assumption that the speed of light in a vacuum was constant, and the Universe was static, neither expanding nor contracting. He also believed that the Universe had no beginning and would never end. His name was Albert Einstein. His theories were called Special and General Relativity. Since these theories did not exactly explain how or why the Universe was not collapsing due to the force of gravity, Einstein added the "Cosmological Constant". This was the antigravity force that opposed gravity and kept the Universe from collapsing in on itself. When Edwin Hubble proved that the Universe was actually expanding, the "Cosmological Constant" and the belief in gravity were abandoned. Einstein adopted the representation "space-time" to replace both gravity and antigravity. These theories have changed the way that many scientists view reality because the mathematics involved are quite elegant, but only if the basic assumptions are true. In order to come to an agreement on any theory, we must first examine and agree upon the theory's basic assumptions.

Many scientists use the representation space-time when referring to the forces that keep the Universe from flying apart or collapsing in on itself. This is important because, if Einstein's theories are to be true, then not only must the speed of light be constant, but space-time must be some type of substance that controls the transmission of light waves and other types of radiant energy, and also directs the forces that cause the movement of everything throughout the Universe. However space-time is a representation that has no definition and does not correspond to any fact of experience, thus it has no meaning in the real world.

Many scientists seem to have substituted the term space-time for the outdated term "ether". In the old theory of physics (now abandoned) ether was a hypothetical invisible substance that was thought to occupy all of space and served as the medium for the transmission of light waves and other forms of radiant energy. It could not be detected by any device or sensation. It had no mass or substance and offered no resistance to anything passing through it. The idea of ether was abandoned because it obviously did not exist, and space-time has basically the same characteristics as ether.

"The modern world began on 29 May 1919."[7] These words are the opening statement of historian Paul Johnson's historical volume *Modern Times*. On that day, photographs of a solar eclipse were interpreted as showing that the light from distant stars does not move in a straight line, it bends when it passes close to the sun. These photographs supposedly confirmed Albert Einstein's concept of curved space and time dilution, thus proving his general theory of relativity. Einstein did not believe that gravity existed, but that heavy objects curved the fabric of space-time, and this curving of space-time is what caused the light to bend in the eclipse photographs.

Time dilution is the theory that the speed of light is constant and time slows down or speeds up depending on the location of an object, in relation to a massive object such as the earth or the sun. Simply stated, the closer an object is to a massive object the slower time is moving for that object. Conversely, the farther away from a massive object a thing is, the faster time moves for that object. Time dilation also explains why an object farther away from a nearby object, not only appears smaller than it actually is, but the farther away object actually is smaller.

The widespread publicity about the eclipse photographs convinced many of the intellectuals of that day that Einstein's theories had been proven. Physicist and philosopher Alfred North Whitehead stated that it was the first time in nearly 300 years that Newtonian physics had been effectively challenged.

Scientists were not alone in sensing the immense significance of Einstein's theory; the general population was also profoundly affected. Few people, including many science teachers, had any clear idea of the scientific content of the Relativity Theory, but the idea of relativity itself struck a responsive chord in a society already leaning toward moral relativism, and questioning traditional absolute truths.

If Einstein's theory rejected Newtonian concepts of absolute time and space, what did that imply about absolutes in morality and religion? As Johnson explains: "Mistakenly but perhaps inevitably, relativity became confused with relativism. It formed a knife to help cut society adrift from its traditional moorings in the faith and morals of Judeo-Christian culture."[8] We believe that the confusion was in the mind of Einstein. His theories are in the form of mathematical equations, and in his own words, "the only trustworthy source of truth", but they are still only theories about mathematical equations.

Isaac Newton believed that the universe was infinite and time had no end, but that the velocity of light was relative. But the universe is not infinite, and time had a beginning and will have an end. But Einstein turned it around: he said the velocity of light is absolute, and space and time are relative. He then derived new transformation laws (mathematical formulas) to replace Galileo's law of addition of velocities. The rest of the Relativity Theory is a mathematical deduction from those laws.

In fact, Einstein's theory is so rigidly deductive that over time it led him away from an empirical view of science (relying or based solely on experiment and observation) to a highly rationalistic approach toward science (relying on mathematical equations for truth and ignoring experiment and observation). Physicist and historian Gerald Holton notes that over his lifetime, Einstein moved away from an early empiricism to a strong emphasis on the mathematical and deductive side of science. Einstein himself writes that during the course of his work he abandoned "skeptical empiricism and became a believing rationalist, that is, one who seeks the only trustworthy source of truth in mathematical simplicity."[8]

Empiricism or inductive reasoning is the experimental method. It is reasoning from particular facts or individual cases to a general conclusion, the search for knowledge by observation and experiment. The logical error committed by Einstein was his belief that science is deductive rather than inductive. Deductive reasoning ignores physical evidence altogether and attempts to force science to agree with what it believes to be true about the Universe. Thus the very nature of science is eroded. Deductive reasoning inhibits us from freely examining physical evidence and drawing unbiased conclusions.

This strong belief in mathematical truth explains an otherwise puzzling story. Einstein, it seems, was only mildly interested in the eclipse that dazzled the rest of the world and provided such spectacular support for his theory. A few years earlier, he had written a personal letter saying he was "fully satisfied with his theory of relativity on purely mathematical grounds—so much so that he wrote—that I do not doubt any more the correctness of the whole system, *may the observation of the eclipse succeed or not.*"[8]

Even when the results of the eclipse were announced, Einstein retained his serenity. A student who was with him that day records that he responded quite calmly to the news, stating he already knew the theory was correct even without this empirical confirmation. The student asked Einstein what he would have done had the results not confirmed his prediction. "Then I would have been sorry for the dear Lord, he replied. The theory is correct."[8]

His turn toward rationalism is perhaps best expressed in his religious views. He wrote, "I believe in Spinoza's God, who reveals himself in the orderly harmony of what exists, not in a God who concerns himself with fates and actions of human beings."[8] Baruch Spinoza was a highly rationalistic philosopher living in Holland in the 1600s. For Spinoza, God was not a Being distinct from the Universe, a personal Creator who brought the world into existence. Instead "God" was merely a name for the principle of order and natural laws that we observe within the Universe.

Einstein did sometimes speak of God as a distinct Being, yet he made it clear that in his view God was completely bound by rational necessity. Einstein wrote: "God Himself *could not have* arranged those connections (expressed in scientific laws) in any other way than that which factually exists."[8] In other words, God had no choice; the laws of science reveal the only possible way He could create the world. He obviously did not believe that God had created the "laws of science".

In short, "Einstein's theory of relativity is simple, elegant, and *mathematically* consistent."[8] However there is another aspect to this story that casts serious doubt on the validity of his basic assumptions. The "rest of the story" is taken from Stephen Hawking's book, *A Brief History of Time*.

"It is normally very difficult to see this effect, (the bending of light rays by gravity) because the light from the sun makes it impossible to observe stars that appear near to the sun in the sky. However, it is possible to do so during an eclipse of the sun, when the sun's light is blocked out by the moon. Einstein's prediction of light deflection could not be tested immediately in 1915, because the First World war was in progress, and it was not until 1919 that a British expedition, observing an eclipse from West Africa, showed that light was indeed deflected by the sun, just as predicted by the theory. This proof of a German theory by British scientists was hailed as a great act of reconciliation between the two countries after the war. It is ironic, therefore, that later examination of the photographs taken on that expedition showed the errors were as great as the effect they were trying to measure. Their measurement had been sheer luck, or a case of knowing the result they wanted to get, not an uncommon occurrence in science."[9] Is this an illustration of the old saying, "figures don't lie but liars do figure"?

The photographs of the eclipse demonstrated that some type of force caused the light from the star to bend toward the sun, and some other force prevented the light from being pulled into the sun. The photographs also demonstrate that the light from the star had to travel farther to go around the sun than it would have if it had traveled in a straight line. Even if the arc the light had followed were only one degree on the surface of the sun, it would have traveled about 747,000 miles farther than if it had traveled in a straight line. It would also have required approximately 4 seconds longer to complete the trip if it was moving at 186,000 miles per second. In reviewing these considerations we can conclude that the photographs only demonstrated that light is subject to forces other than its own inertia.

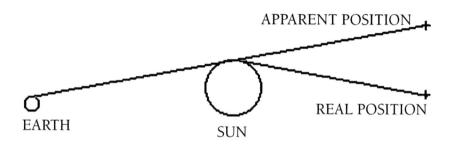

This figure is commonly used to show what happens to light when it comes near a high gravity body; it is supposed to demonstrate the validity of Einstein's theories. If it were closer to scale, it would be easier to see just how far the light was bent and how much farther it had to travel to get around the sun. This diagram does shows that if an object travels at the same velocity from one point to another it will take longer to go around a curve than to travel in a straight line.

There are a number of other problems when Einstein's absolute is applied to the real world, in addition to the photographs of the eclipse. For example, it is not possible to measure the velocity of light in a vacuum, because a perfect vacuum does not exist in this universe (or at least we have not yet found or created one).

Time dilation is only an optical illusion, a lack of depth perception, or a misunderstanding of perspective, and is illustrated by the following photographs.

Photo 1

Photo 2

Photo 3

These photographs were taken with a single lens, Polaroid One Step Auto Focus camera, using Polaroid 600 platinum film. The model is a board that is 15 inches long, 5 3/4 inches wide, and 3/4 of an inch thick. The dowel rods are 5/16 of an inch in diameter and 2 3/8 inches high. The one stripe and four stripe rods are each 1 1/2 inches in from their respective ends of the board and there is a 4-inch space between the rods.

Photo 1. Shows that if the model is turned so that the one stripe rod is closest to the observer and the other rods are placed so that the two stripe rod is next farthest away, the three stripe rod is next farthest away, and the four stripe rod is farthest away from the observer; the observer will perceive the closest rod to be the largest and the farthest rod as being smallest.

Photo 2. Shows that a human observer with normal depth perception, will observe the same effect if the model is reversed.

Photo 3. Shows that when a two dimensional picture is taken and the objects of attention (the striped rods) are the same distance away from the observer (the camera) and since they are the same size in the real world, they will be, or appear to be, the same size in a two dimensional world (the picture).

This illustrates that the objects farther away but of equal size, will always appear to be smaller in a two dimensional frame of reference. Thus it is not a matter of true size but of perception and perspective. Believing that mathematical equations are the only source of truth can result in rather bizarre conclusions about the real world.

The human visual system is much more complicated than a camera, even more than a high-speed movie camera. The combination of two eyes and two visual centers in the brain allows for stereopsis (the ability to see things in three dimensions). If necessary (in cases of brain damage or loss of vision in one eye), the brain can still use motion parallax, (the ability to detect changing distance, with both eyes or with only one eye). This is achieved by correlating the impression of changing size of the observed object and motion of the background. The idea of objects being different sizes in different frames of reference only applies to two-dimensional worlds. As you know by observation, this earth and the entire universe exist in at least three dimensions.

Another erroneous assumption, based on Einstein's belief that time and space are relative, is that a person needs an outside frame of reference to determine motion. This assumption is based on the following illustration. "The central effect of Einstein's theory was to make time and space relative. What does that mean? Imagine yourself sitting in a subway train, looking across the tracks at another train, when yours begins to move. But wait—is it really your train moving? Or is it the other one? If the train accelerates gently enough, you may not be sure for a moment."

"Nor can you answer the question simply by looking at the trains themselves. By looking across at the other train, all you can determine is that they are moving relative to each other (relative motion). To determine which train is "really" moving, you have to look at the station platform. The train that's "really" moving is the one leaving the station. Yet as Einstein might point out, the station itself is also moving. It's anchored to the earth, to be sure, but remember that the earth is moving relative to the sun. And the sun is moving relative to the Milky Way galaxy. And the galaxy is moving relative to the so-called fixed stars (though they are not really fixed either). In short, all motion is relative to some other reference point. Where is a frame of reference that is absolutely at rest?"

"There is none Einstein said."

"To make the point clearer, consider another scenario. Imagine that you are flying in a spaceship out in empty space, and you see another spaceship approaching. But wait—is it really approaching you or are you approaching it? Given that both ships are in uniform motion, you cannot answer that question. (Uniform motion is a technical term meaning motion in a straight line with constant speed.) You can observe that the distance between the two spaceships is decreasing, but you cannot tell whether only one ship is moving or if they both are. There is no celestial subway platform to use as a reference point."[10]

Both of these scenarios are supposed to illustrate the expansion or contraction of time and space and thus prove Einstein's theories. But remember they are both imaginary illustrations. They both assume that the only way a human can detect motion is by using the sense of sight. What happens if the train in the subway station has no windows or the observer in the train is blind? The Creator placed a backup system in the human nervous system for just such a problem. This backup system consists of the vestibular apparatus in the inner ear and the position and motion sensors in the muscles and joints. A blind person routinely uses this system. When there is confusion in the visual apparatus the sighted person can determine motion and direction of motion by ignoring the visual input, and relying on the backup system.

The two-spaceship scenario is no longer imaginary. The Space Shuttle pilots are faced with the problem every time they have to dock with a space station or another object in space. The space pilot must know in what direction his (or her) ship is moving and in what direction the dock is moving. He must position his ship so that it will overtake the dock and then he must slow down enough to match the velocity and direction of the dock at the moment the two vehicles touch. This is not an imaginary scenario, and does not require the expansion or contraction of time. It does require the skills of an intelligent human and the use of machines that have been constructed by intelligent humans.

Both Newton and Einstein made wrong assumption in order to prove their theories. Newton believed that space and time were created things and were absolutes. Einstein believed that space and time were real things but were relative and the only absolutes were the velocity of light in a vacuum, and his equations. In the real universe space is not a thing, and "the fabric of space-time" cannot be detected by any means known to man. Space is the empty area between bits of matter. Time is not a thing. It is the period from the beginning to the end of an event.

Recent experiments may have made untenable the whole idea of time dilatation, curved space, and objects being different sizes in different frames of reference. "Despite some recent virtuosic experiments with pulses of light widely reported to far exceed the speed of light, physicists still agree that no object or information has been made to travel superluminally (faster than the speed of light) . . . Creating the most recent hubbub is a clever experiment in which a pulse of light propagates superluminally through a cell of cesium gas. The group velocity (the velocity of a pulse undistorted in shape) is negative, a counterintuitive situation that means the peak of the pulse arrives at the end of the cell in a time that is less than that of an equivalent pulse traveling through a vacuum. In fact, because the group velocity is negative, it exits the cell even before it enters it."[11] The pulse actually traveled through the cesium cell 310 times faster than a similar pulse would travel through a similar sized vacuum cell. Since Einstein's theory of relativity depends on the statement that "nothing can travel faster than the speed of light in a vacuum", the scientists at the NEC Research Institute in Princeton, N.J. seem to have proven that the theory is not correct.

The representation space-time has now been replaced by the representation "dark energy", the force that is causing the Universe to expand. Scientists are not certain just what this energy or force is. However if each of the innumerable stars in the Universe is almost continuously fusing hydrogen into helium, there is an enormous amount of force being distributed in all directions throughout the Universe. Could this be the dark energy that is causing the expansion of the Universe? If $E=mC^2$, there is plenty of force available in the stars.

Another mistaken assumption is that the formula $F=ma$ (force equals mass times acceleration) is a mathematical equation. It is actually an engineering formula to determine how much force must be applied to a given mass to "force" it to increase or decrease its velocity. Force in this sense is applied to the mass to push it, slow it down, change its direction or stop it motion. An example would be the launch of a spacecraft. A force is applied to the craft to accelerate it form zero velocity to orbital velocity but some of the mass of the fuel is changed into force and used up producing the acceleration. The mass of the craft actually gets smaller as it accelerates not larger. What changes is its momentum, and inertia. Force is always a push that is applied to a mass, but not added to the mass. Force can either be positive or negative, producing either acceleration or deceleration, depending on the direction in which the force is applied to the mass. As stated in the section on communication, "a representation that cannot be linked to a sensory experience is a representation that has no meaning." This applies to mathematical equations as well as written or spoken "representations".

The discussion about frames of reference and two and three-dimensional worlds is to illustrate that time is not a property of this universe. It is the continuum through which this universe is moving. There are other illustrations of how human physiology, if functioning normally, will let you know whether you are accelerating, decelerating, or changing direction. These abilities are build into the human body in the inner ear, in the position sensors in the muscles and joints, and in the vision centers in the optical areas of the brain. Inputs from these areas are then coordinated in the cerebellum part of the brain. Attempts to define reality, in a three dimensional universe, are doomed to failure if things are viewed as only two-dimensional.

To rely on mathematical equations as final reality, instead of a Creator who is unchanging is both foolish and dangerous. Only God is a fixed reference point, and never changes.

> Genesis 3:13–14: "Then Moses said to God, 'If I go to the Israelites and tell them that the God of their forefathers has sent me to them, and they ask me, "What is His name?" what shall I say?' God answered, 'I AM; that is who I am. Tell them that I AM has sent you to them."

> Hebrews 13:8: "Jesus Christ is the same yesterday, today, and forever."

CHAPTER 8

The Bible, Truth Or Myth?

Psalm 14:1 & 53:1: "**The impious fool says in his heart, there is no God.**"

Satan knew that even if educators became "fools", i.e. stopped believing that God exists, many people would continue to believe that the Bible is a source of Truth. Satan knew that challenging the veracity, authenticity, and authority of the Bible, would be one of the most effective means of eroding the basis for man's belief in God. This would be particularly effective if the questions about the truth of the Bible came from within the church itself.

In the late 19th century liberal theologians began to question the identity of the authors of the individual books of the Bible. They even claimed that the Gospel writers had a common source text, which they copied. These "scholars" refer to this mysterious source as the Q document. Although it has never been found, nor have any historical references to its existence been identified, these theologians continue to insist that it must exist. Such scholars are guilty of making up falsehoods to fit their own ideas of what they think the Bible should really be. These theologians are making up fairy tales; they are more foolish than the public school instructors who are teaching that life occurred by random chance.

Today, many churches, including several mainstream denominations, have adopted these heresies in an attempt to conform to our society's desires to be tolerant or politically correct. God does not change, nor does truth. They are both absolutes. However these churches have ignored the words of the Bible.

> Revelation 22:18&19: "I warn everyone who hears the words of the prophecy of this book. If anyone adds anything to them, God will add to him the plagues described in this book. And if anyone takes words away from this book of prophecy, God will take away from him his share in the tree of life and in the holy city, which are described in this book."

> Hebrews 6:4–6: "For when men have once been enlightened, when they have had a taste of the heavenly gift and a share in the Holy Spirit, when they have experienced the goodness of God's word and the spiritual energies of the age to come, and after all this have fallen away, it is impossible to bring them again to repentance; for with their own hands they are crucifying again the Son of God and making mock of His death."

It is important to realize that much of this decay of the church is happening from the inside. Church members, for their own convenience, are changing Scripture and embracing beliefs that are contrary to the Bible. Satan is more successful when he attacks from within the church family than by overt attacks on Christians. Now many Christian churches have forgotten what Paul told the church at Corinth.

> 1 Corinthians 2:1&2: "As for me, brothers, when I came to you, I declared the attested truth of God without display of

fine words or wisdom. I resolved that while I was with you I would think of nothing but Jesus Christ; Christ nailed to the cross."

1 Corinthians 1:18: "This doctrine of the cross is sheer folly to those on their way to ruin, but to us who are on the way to salvation it is the power of God."

Once church leaders began to doubt the authority of the Bible, Satan knew that it would be easier to undermine man's belief in the existence of God. If the idea of a creation were eliminated, it would be much easier to deny the existence of a Creator. Educators, from the elementary to the postgraduate level, would accept random chance as truth and not merely a theory. Ultimately the theory of evolution became required teaching in almost all schools and universities in this country.

It was necessary to get people to believe the lie of random chance in order to make the idea of a Creator an abstract choice instead of a reality. If there is no Creator then nothing is absolute! God does not exist. If there is no God, then there is no right or wrong. If right or wrong is only relative then there is no morality. Values, such as truth, honor, love, family, duty, and even life itself are meaningless. How did Hitler justify killing six million Jews? With no God, truth is whatever you want it to be.

Apart from the Creator God, our nation and our government have no basis for existence. Read the words of the Declaration of Independence. "When in the Course of human events it becomes necessary for one people to dissolve the political bands which have connected them with another, and to assume among the Powers of the earth, the separate and equal station to which the <u>Laws of Nature and of Nature's God</u> entitle them, a decent respect to the opinions of mankind requires that they should declare the causes which impel them to the separation. <u>We hold these truths to be self-evi-</u>

dent, that all men are created equal, that they are endowed by their Creator with certain unalienable Rights, that among them are Life, liberty, and the pursuit of Happiness."

The US Supreme Court has ruled that we cannot even mention God, the Creator, in our public schools. Theoretically we are no longer allowed to teach the Declaration of Independence in the public schools. Logic then implies that there are no such things as "unalienable rights"! We are now living in a nation where students in our public schools are being taught that there are no absolutes (no laws), nothing is right or wrong. If there is no Creator and no right or wrong, then the only morality is what feels good at the time. If this is true, and our government's leaders have been acting as though it is true, then they can do anything they desire. The Supreme Court can say that the Constitution is not what is written, but rather is what the justices interpret it to be!

If there is no Creator, then the Bible is just a collection of myths and legends. This Humanist type of reasoning provides no basis for the Ten Commandments or any Divine Law. In 1963 the United States Supreme Court openly stated that the Bible and Divine Law were no longer the basis for our laws. From that point forward "the law" has become whatever the Court wants it to be. And what have the results been? We can observe the decay and destruction of our society subsequent to this shift in worldview.

By 1995, in the United States, violent crime had increased from 250,000 per year to over 1,700,000 per year. Pregnancies in unwed girls under 15 years of age had increased from 4,000 per year to over 26,000 per year. Sexually transmitted diseases had increased from 400,000 cases per year to over 1,000,000 cases annually. Aids became an uncontrolled epidemic. Drug abuse had become prevalent even in elementary students. As Ben Franklin predicted, "We have been divided by our partial local interests; our projects have been confounded, and we ourselves have become a reproach and bye word to the rest of the world."[12]

We cannot have it both ways. We must choose, and our response to everything in life depends on that choice. Either there is a Creator, this is a created universe where scientific methodology works, morality exists, the Bible is the authentic written Word of God, and the Ten Commandments are absolute and the basis for all law; or everything is the result of random chance and nothing matters. The choice is ours. How we live will depend on the choice we make.

CHAPTER 9

Admissible Evidence

What do you believe about the Bible? Do you believe that it is simply a religious book, or a collection of myths, or a true account of historical events? In the preceding chapters of this book we have quoted a number of statements from the Bible. Why do we believe that these statements are important? We believe that the Bible is important because it is one of the three fundamental bases on which the Christian worldview stands. They may be important for believing the Christian worldview, but are they true?

What evidence do we have that the Bible is true? Why should we believe it and not other religious books? In July 1870 the Vatican Council issued a proclamation that established the doctrine of Papal Infallibility. The proclamation means that when the Pope speaks ex-cathedra, by virtue of his office, he is speaking the very words of God. Therefore the Pope's words are free from any error.

Since the time of Martin Luther the Protestants have taught that the Bible is truth because men who were God's servants wrote it. The Fundamentalists carried this one step further and stated that the Bible was verbally inspired. Verbal inspiration means that the writers of the Bible were simply writing the words that God had dictated to them. Today this teaching is called inerrancy but has a slightly different meaning. Inerrancy means that when the authors of the original documents wrote these manuscripts, God did not allow them to make any errors.

In an essay entitled *Inerrancy or Verbal Inspiration? An Evangelical Dilemma*, J. Ramsey Michaels points out that questions of inspiration are matters of belief. He then goes on to say, "It has been said to me: 'On that same basis someone could equally well accept the *Koran*, or the *Book of Mormon!*' The appropriate answer is, 'Of course he could. After all, millions do.' But the subject of inerrancy or verbal inspiration has not been the question of why one should choose Christianity over other options, but how those who are already Christians should regard the Bible."[13]

Any one or all of the ideas about inspiration may be true, but they are all axioms and cannot be used to prove that the Bible is true. An axiom is a statement that needs no proof, and is simply accepted as factual. This does not mean that any or all of these ideas are false, only that they cannot be proven to be true. Such ideas are useful for theological discussions, but are not helpful in proving the truthfulness of the Bible.

The correct way to prove to the non-Christian that the Bible is true and that the God of the Bible is the Creator, will be shown in the following discussion of what constitutes acceptable legal testimony and where the burden of proof for the truth or falsehood of such testimony resides?

Robert Gange, President of The Genesis Foundation, put it this way; "There are only three categories of events in space and time: events that are reproducible, events that are unpredictable, and events that are singular. An event that is reproducible is the data of science. If you, for example, want to measure how fast an apple hits the ground when released a certain height above the ground, you can repeat that over and over and over again. So reproducible events lend themselves to scientific inquiries, whereas unpredictable events lend themselves to statistical inquiries. Events that are singular lend themselves to legal inquiries. And the questions of whether there was a creation of the world or there is a destiny to man are singular inquiries. These are singular events. The creation of the world was a one-time event, and as such, it is a legal inquiry. Science has no proper jurisdiction in the question of origin or destiny."

Like creation, much of the Bible is about singular events. Thus inquiry into the veracity, authenticity, and authority of the Bible is a legal inquiry. Some of what is written in the Bible was not observed, and some has not yet happened. How can we know if any of it is true? If we can show, by legal inquiry, that even one of the most unlikely stories in the Bible is true, it is easy to conclude that all of the stories are true, because the same legal arguments apply to all written testimony. The most unlikely event that is recorded in the Bible is that a man who was undeniably dead, buried, and had been in the grave for at least part of three days, came back to life.

There are two such stories in the Bible. Stories of men, who died, were buried, and remained in the grave for part of three days or longer. Then they lived again and were seen by many witnesses. Because these stories are written testimony, the rules of legal evidence apply when attempting to authenticate them. The following legal arguments are taken from Simon Greenleaf's book *The Testimony of the Evangelists*.

Simon Greenleaf (1783–1853) was born in Newburyport, Massachusetts. He studied law and was admitted to the bar in 1806. He began his practice of law in Portland, Maine and served as chief reporter for the Maine Supreme Court for twelve years. He then accepted the Royall Professorship of Law at Harvard Law School.

He held this position until his death in 1853. His three-volume work, *A Treatise on the Law of Evidence*, is considered a classic of American jurisprudence. It is also the greatest single authority on the Common Law rules of evidence in the entire literature of legal procedure. The section on the rules for admissibility of written testimony forms the basis for *The Testimony of the Evangelists.* These rules are just as valid today as they were when he first wrote them and are the rules routinely used in courtrooms where Common Law is practiced.

Professor Greenleaf wrote *The Testimony of the Evangelists* for lawyers. Kregel Classics, Kregel, Inc., P.O. box 2607, Grand Rapids MI 49501, published an edition in 1995. It should be available from any of the large Internet bookstores. The following quotations are from the 1874 edition.

To the Members of the Legal Profession

Gentlemen,

"The subject of the following work I hope will not be deemed so foreign to our professional pursuits, as to render it improper for me to dedicate it, as I now respectfully do, to you. If a close examination of the evidence of Christianity may be expected of one class of men more than another, it would seem incumbent on us, who make the laws of evidence one of our peculiar studies. Our profession leads us to explore the mazes of falsehood, to detect its artifices, to pierce its thickest veils, to follow and expose its sophistries, to compare the statements of different witnesses with severity, to discover truth and separate it from error. Our fellow men are well aware of this; and probably they act upon this knowledge more generally, and with a more profound repose, than we are in the habit of considering. The influence, too, of the legal profession upon the community is unquestionably great; conversant, as it daily is, with all classes and grades of men, in their domestic and social relations, and in all the affairs of life, from the cradle to the grave. This influence we are constantly exerting for good or ill; and hence, to refuse to acquaint ourselves with the evidences of the Christian religion, or to act as though, having fully examined, we lightly esteem them, is to assume an appalling responsibility."

"The things related by the Evangelists are certainly of the most momentous character, affecting the principles of our conduct here, and our happiness forever. The religion of Jesus Christ aims at nothing less than the utter overthrow of all other systems of religion in the world; denouncing them as inadequate to the wants of men, false in their foundations, and dangerous in their tendency. It not only solicits the grave attention of all, to whom its doctrines are presented, but it demands their cordial belief, as a matter of vital concernment. These are no ordinary claims; and it seems hardly possible for a rational being to regard them with even a subdued interest; much less to treat them with mere indifference and contempt. If not true, they are little else than the pretensions of a bold imposture, which, not satisfied with having already enslaved millions of the human race, seems to continue its encroachments upon human liberty, until all nations shall be subjugated under its iron rule. But if they are well-founded and just, they can be no less than the high requirements of heaven, addressed by the voice of God to the reason and understanding of man, concerning things deeply affecting his relations to his sovereign, and essential to the formation of his character and of course to his destiny, both for this life and for the life to come. Such was the estimate taken of religion, even the religion of pagan Rome, by one of the greatest lawyers of antiquity, (Cicero) when he argued that it was either everything or was nothing at all. *Aut undique religionem tolle, aut usquequaque conserva.*"

"With this view of the importance of the subject, and in the hope that the present work may in some degree aid or at least incite others to a more successful pursuit of this interesting study, it is submitted to your kind regard, by your obedient servant, Simon Greenleaf"[14]

The summary following each rule is a very short paraphrase of the text of Professor Greenleaf's book. For those who want to know and use the full arguments for each rule, we suggest that you obtain a copy of the 1995 edition of *The Testimony of the Evangelists*.

The testimony being considered is the written accounts, entitled the Gospels, which are included in the New Testament portion of the Bible. Each Gospel is identified by the name of the writer, but no original manuscripts of these Gospels exist today.

The first rule of evidence concerning ancient writings:

"Every document, apparently ancient, coming from the proper repository or custody, and bearing on its face no evident marks of forgery, the law presumes to be genuine, and devolves on the opposing party the burden of proving it to be otherwise."[15]

In as simple English as possible, the first rule to prove the genuineness of any old document, such as the Bible, is that it must be found in the place where, and in the care of persons, with whom one might naturally and reasonably expect to find these types of documents. The Bible is both an historical and a religious book. Thus a reasonable person would expect to find the Bible in public libraries, where it is available for public use as an historical reference, and in religious institutions, where it is used as a source of religious teaching. The Bible is routinely found in these places and in many other places as well, it passes the first requirement for genuineness.

The second requirement for genuineness is that the document does not appear to be a forgery. There is no pretense that the Bible was written on golden plates and discovered in a hidden cave, or was brought to this world by some angelic being. The Bible has always been accepted as the ordinary writings of the men whose names appear in the documents. These documents were made public when they were first written, and have remained as public documents ever since. It is not important that only copies are available now, and that no effort was made to preserve the originals, because the making of copies was public knowledge and was done under the direction and guidance of the proper custodians of the documents. It is important to note that no objections have been made to the accuracy of the copies at any time, then or now.

The second rule of evidence concerning ancient writings applies to the accuracy of copied documents:

> *"In matters of public and general interest, all persons must be presumed to be conversant, on the principle that individuals are presumed to be conversant with their own affairs."*[16]

Since there is no argument about the proper custody of the Bible or about the accuracy of the copies now in existence, and there is no suggestion that the Bible is a forgery, the Bible is acceptable as written testimony that the writers believed to be true. The person, who accepts the Bible as genuine and authentic, does not need any other proof than this argument. By the plainest rules of law, the one who does not accept the Bible as genuine and authentic must prove his case; the burden of proof lies with him.

These same two rules apply to properly authorized translations i.e. translations by proper custodians and no evidence of forgery.

Therefore, we can accept copies of the Bible and appropriately authorized translations as being authentic, but why should we believe that the testimony of these witnesses is true? The copies may be authentic copies, and the translations may be authentic translations, but how do we know that the testimony of the writers of the Bible is true?

The legal tests for the truth of testimony:

"In trials of fact, by oral or written testimony, the proper inquiry is not whether it is possible that the testimony may be false, but whether there is sufficient probability that it is true." [17]

"A proposition of fact is proved, when its truth is established by competent and satisfactory evidence." [18]

"In the absence of circumstances which generate suspicion, every witness is to be presumed credible, until the contrary is shown; the burden of impeaching his credibility lying on the objector." [19]

"The credit due to the testimony of witnesses depends upon, firstly, their honesty; secondly, their ability; thirdly, their number and the consistency of their testimony; fourthly, the conformity of their testimony with experience; and fifthly, the coincidence of their testimony with collateral circumstances." [20]

The first three statements are the assumptions that the court makes about testimony. The fourth statement is the tests that are applied to the witnesses.

First, the honesty of the witnesses: The rule of law is that ordinary men will tell the truth when they have nothing to gain by lying. The major truths that these witnesses adhered to were that Jesus had died, and then come back to life, and that only through repentance from sin, and belief in Jesus and His resurrection from the dead, could anyone have any hope of life after death. These beliefs they all agreed to and were unshakable in their insistence that they had actually seen Jesus truly alive after they had seen Him truly dead.

If they were not absolutely certain that they were telling the truth, they had every reason to be quiet or to admit that the story was a lie. Jesus had just been executed as a common criminal, after being sentenced in a public trial. His religion would overthrow every other religion in the whole world. All of the rulers and great men of the world were against them. The general public was against them. If they persisted in their testimony they could expect nothing but contempt, opposition, bitter persecution, beatings, jail, torture and death! It is impossible that they could have persisted in their assertion that Jesus was alive, if they had not known it to be fact as sure as any other fact that they knew.

To continue in a lie that they knew to be a lie cannot be reconciled with the fact that they were common ordinary men. All of the rest of their testimony shows them to be common men like all other common men. Even the members of the Sanhedrin, the Supreme Court of Israel, recognized that they were common men. They had the same motives, the same hopes, the same joys, sorrows, tears, passions, temptations, and the same weaknesses as we have. Their writings also show them to be men of sanity and intelligence. There was no possible motivation for them to lie.

If they had been bad men then they might have lied just to be lying, but it is nonsense to suppose that bad men would invent lies to promote the religion of the God of Truth.

Were these witnesses honest? The only rational answer is yes. They were honest men, testifying to what they had carefully observed and considered, and knew with certainty to be true.

Secondly, the ability of the witnesses: The rule of law is that the ability of a witness to speak the truth depends on the opportunities that he has had for observing the fact, the accuracy of his powers of observation, and the faithfulness of his memory in retaining the facts once they were observed. Since we have no real knowledge of their powers of observation or the faithfulness of their memories, the law allows us to presume that these men were honest, of sound mind, and possessed of the average and ordinary degree of intelligence. This presumption is always made unless there is an objector who can prove that the presumption is not true. In addition, Matthew, as a tax collector, was trained to be suspicious of any unlikely story; and Luke, as a physician, was trained to be a careful observer of any events concerning life and death. In the absence of any objection we may safely assume that these were men of at least ordinary ability.

Thirdly, the number of the witnesses and the consistency of their testimony: The character of the written testimony contained in the Bible is like that of all other true witnesses, that is it contains substantial agreement about a variety of circumstances, and enough difference to show that the writers <u>did not compare notes or collaborate in their testimony.</u> To think that they might have conspired together to lie to the world is inconsistent with the supposition that they were honest men, a fact that has already been demonstrated. If they were bad men who had conspired to lie, there would have been no differences in their testimonies. Since collusion or conspiracy to lie has been excluded, the only explanation for the consistency in their testimony is that it is true. It is beyond reasonable doubt to think that four men could tell such an improbable story as a man coming to life after being dead, and agree on so many particulars in their stories, without conspiring together, unless the story was true.

Fourthly, the conformity of their testimony with experience: The testimony that a man who was truly dead, had been in the grave under guard for at least part of three successive days, and had then come back to life is certainly contrary to human experience. However each individual fact about the life, trial, execution, burial, and appearances of Jesus after his death, if taken individually, is not contrary to human experience.

There is no doubt that Jesus was a real historical person. His trial was conducted in the open, in front of a large crowd of people. His execution was conducted in a public place, in the open countryside, in front of many witnesses. His death was certified to the satisfaction of Roman and Jewish officials, by His executioners. His burial was supervised by a group of soldiers who had been appointed by his executioners. After He had come back to life, He appeared to over five hundred individuals, some singly, but usually in groups. He talked with them, ate with them, showed them His wounds, and insisted that they touch Him to be sure that He was really alive. Several witnesses were told that Jesus was alive, but they did not believe it until they had seen Him themselves and observed that He was truly alive.

Each of these events taken by itself and testified to by such a variety of witnesses would be sufficient evidence to satisfy any court as to the truth of the event, even one as unlikely as a dead man coming back to life.

The testimony of two of the witnesses, Peter and John, convinced the Sanhedrin, the Supreme Court of Israel, that Jesus had come back to life. When these men testified to this fact before the court, instead of punishing them for lying, the court instructed them not to tell anyone else that Jesus was alive. The members of the Sanhedrin did not want to believe Peter and John, but the evidence was so compelling that they had no other choice.

Fifth and finally, the coincidence of their testimony with collateral circumstances: This means, how much extra information do the witnesses give that is unrelated to the facts under investigation. In other words, how much detail is included in the testimony that can be shown to be true or false from other contemporary sources, not associated with the witnesses. For example; if the story of the trial, execution, burial, and resurrection of Jesus were not true, the names and offices of the members of the Sanhedrin, the Roman officials, the man who supervised the burial, the exact time of day that the execution occurred, and numerous other incidental events would not have been mentioned. Names, places, and times are too easy to verify. So false witnesses will rarely mention them.

The names of many of the people involved in the Biblical accounts have been verified by a number of secular historians. The Roman historian, Tacitus, and the Jewish historian, Josephus, both name Pontius Pilate as the Roman governor in Judea at the time of Jesus trial. The evangelists tell us that the accusation against Jesus was written in three languages and posted on His Cross. The Roman historian, Suetonius tells us that this posting was commonly done at executions. Josephus mentions that public notices in and around Jerusalem were commonly written in three languages. Many of the events described by the evangelists have also been described by secular historians, such as the removal of the prisoner's garments and whipping before the execution. To list all of these verifications would take a textbook of many hundreds of pages. However there are numerous books of history and archeology that are available in many public libraries that verify the names and events mentioned in the Bible. The National Geographic Society has published a book, *Every Day Life in Bible Times*. This entire book is a verification of the names of people, places, and events as recorded in both the Old and New Testament. There is even a map included with the book that shows the present day location of all of the places mentioned by Matthew, Mark, Luke, and John.

Considering the volume of evidence available that collaborates the testimony of the evangelists, an honest person must conclude that they were honest men who faithfully reported events as they actually saw them happen. The rules of law and evidence allow no other conclusion.

How do we know that the Bible is true? We have at least four honest men who have provided us with accurate testimony about the great truths of the Bible. Because these same rules of evidence apply to all of the testimony contributed by each of the Biblical writers, we may legally conclude that the entire Bible is true testimony, given by honest men. If an unbeliever wishes to challenge this conclusion, the burden of proof lies with that unbeliever.

CHAPTER 10

Christianity

Final Reality, that which leaves no further chance for action, discussion, or change, is the immortal, personal God, who exists in Eternity, and is not part of the universe, space, or time. He has shaped all matter and energy into its present form by the power of his WORD. This was the Christian worldview, and it had dominated Northern Europe and then the United States for hundreds of years. Then something happened. Humans, Christian and non-Christian, began to listen to Satan again just as they had in Eden. They decided that they wanted to be as smart as God.

> Genesis 3: 1–6: The serpent was craftier than any wild creature that the Lord God had made. He said to the woman, "Is it true that God has forbidden you to eat from any tree in the garden?" The woman answered the serpent, "We may eat the fruit of any tree in the garden, except for the tree in the middle of the garden; God has forbidden us either to eat or to touch the fruit of that tree; if we do we shall die." The serpent said, "Of course you will not die; God knows that as soon as you eat it, your eyes will be opened and you will be like God know-

ing both good and evil." When the woman saw that the fruit of the tree was good to eat, and that it was pleasing to the eye and tempting to contemplate, she took some and ate it. She also gave her husband some and he ate it.

Humans decided that they did not need God any more and they could be their own gods. With the exception of a few Christian schools, Christian colleges, and some home schools, the vast majority of public and private schools and colleges in the United States are now teaching the Humanist worldview of random chance. Francis A. Schaeffer explains the difference between the two worldviews in his book *A Christian Manifesto*.

"The basic problem of the Christians in this country in the last one hundred years or so, in regard to society and in regard to government, is that they have seen things in bits and pieces instead of totals."

"They have very gradually become disturbed over permissiveness, pornography, the public schools, the breakdown of the family, and finally abortion. But they have not seen this as a totality, each thing being a part, a symptom, of a much larger problem. They have failed to see that all of this has come about due to a shift in worldview, that is, through a fundamental change in the overall way people think and view the world and life as a whole. This shift has been *away from* the worldview that Final Reality is a personal God, who has shaped all matter and energy into its present form by the power of his Word. This worldview was at least vaguely Christian in people's memory (even if they were not individually Christian). The shift has been *toward* something completely different, toward a worldview based upon the idea that final reality is impersonal matter or energy shaped into its present form by impersonal chance. They have not seen that this worldview has taken the place of the one that had previously dominated Northern European culture, including the United States, which was at least

Christian in memory, even if the individuals were not individually Christian."

"These two worldviews stand as totals in complete antitheses to each other in content and also in their natural results including sociological and governmental results, and specifically including law."

"It is not that these two world views are different only in how they understand the nature of reality and existence. They also inevitably produce totally different results. The operative word here is *inevitably*. It is not just that they happen to bring forth different results, but it is absolutely *inevitable* that they will bring forth different results."[21]

Both of these worldviews are religious in nature. The Christian worldview is overtly religious; the Bible, and the Universe itself are the main sources for this worldview. It is clearly delineated in three different places in the Bible.

Genesis 1:1: **"In the beginning God created heaven and Earth"**.

John 1:1–3: **"When all things began, the Word already was. The Word dwelt with God, and what God was, the Word was. The Word, then, was with God at the beginning, and through him all things came to be; no single thing was created without him. All that came to be, was alive with his life, and that life was the light of men."**

Romans 1:19&20: **"For all that may be known of God by men lies plain before their eyes; indeed God Himself has disclosed it to them. His invisible attributes, that is to say His everlasting power and deity, have been visible, ever since the world began, to the eye of reason, in the things He has made."**

"Why have the Christians been so slow to understand this (the shift in worldview)? There are various reasons but the central one is a defective view of Christianity. This has its roots in the Pietist movement . . . Pietism began as a healthy protest against formalism and a too abstract Christianity. But it had a deficient, platonic (limited) spirituality. It was platonic in the sense that <u>Pietism made a sharp division between the spiritual and the material world giving little or no importance to the material world. The totality of human existence was not afforded a proper place. In particular it neglected the intellectual dimension of Christianity.</u>"

"Christianity and spirituality were shut up to a small, isolated part of life. The pietistic thinking ignored the totality of reality. The poor side of Pietism and its resulting platonic outlook has really been a tragedy not only in many people's individual lives, but in our total culture."

"True spirituality covers all of reality. The Lordship of Christ covers all of life and all of life equally. It is not only that true spirituality covers all of life it covers all parts of the spectrum of life equally. In this sense there is nothing concerning reality that is not spiritual."[22] I need to emphasize these thoughts. *For the Christian, true spirituality should cover all of reality. There should be nothing concerning reality that is not spiritual.*

It is obvious that many Christians have abandoned the idea that truth about the Universe is really God's truth, and have concentrated on "spiritual" things. They have given the material world to the Humanists, even to the point of accepting random chance as the origin of the material world and life itself.

Webster's Unabridged Dictionary defines logic as "the science of true reason or speech." The dictionary also indicates that logic comes from the Greek word logos. The Greeks believed that logic or true reason was final reality. In the Gospel according to John, he (John) applies this belief to Jesus when he calls Him the Logos. (John could have just as easily called Him Final Reality). Basically Jesus is the way God reasons. He is the WORD, the true speech from God. God used logic, reason, truth, the WORD when He created everything that is. Real Christianity, as illustrated in the Bible, the Bible itself, and the created Universe are truth. They are not just about truth; they are TRUTH. They are reasonable; they are the way God thinks. If we believe that God is real, we would do well to pay attention to what He has to say, and to recognize what He has created.

CHAPTER 11

In The Beginning

Genesis 1:1–2: "In the beginning God created heaven and earth. The earth was without form and void, with darkness over the face of the abyss, and the Spirit of God hovering over the surface of the waters."

This is also written in a slightly different way in John's Gospel in The New Testament.

John 1:1–4: "When all things began, the Word already was. The Word dwelt with God, and what God was, the Word was. The Word, then, was with God at the beginning, and through Him all things came to be: no single thing was created without Him. All that came to be was alive with His life, and that life was the light of men."

This is obviously a statement that has been accepted by faith but there is interesting scientific evidence that this is actually what happened "in the beginning". "The Big Bang theory of cosmology assumes that the universe began from a singular state of infinite density. This theory was implicit in the complete solution of Albert Einstein's equations, obtained by Aleksandr Friedmann in 1922. In 1927, Georges Lemaitre used these equations to devise a cosmological theory incorporating the concept that the universe is expanding from an explosive moment of creation.

"The term Big Bang, as a name for the initial cataclysmic event, was coined (1946) by George Gamow, who with R. A. Alpher envisaged a high temperature state in the beginning and elaborated the theory to include a theory of element synthesis and background radiation. In the light of the evidence currently available, including the discovery of the background radiation, this theory appears to best account for the evolution of the universe."[23]

The term "Big Bang" was, at first, an attempt to ridicule the idea of creation. However with the discovery of "Wrinkles In Time" (phase transitions) in the development of the universe, it became obvious that there really was a beginning. "Big Bang" is still the commonly used name for this beginning, because the theory attempts to explain how force was changed into matter and energy. We know that atomic fusion and atomic fission both release very large amounts of energy and explosive force. We also know that force can be changed into matter and energy, but we do not know how to do this.

The Christian Worldview teaches that God shaped all matter and energy into its present form by the POWER of the WORD. This matter and energy did not come from pre-existing matter or energy. Creation was ex nihilo (out of nothing).

Hebrews 11:3: "By faith we perceive that the universe was fashioned by the Word of God, so that the visible came forth from the invisible." The King James translators put it this way, "Through faith we understand that the worlds were framed by the Word of God, so that things which are seen were not made of things that do appear." In modern day English, "We believe that God created the Universe by the Power of His Word. It did not just happen by random chance."

In 1992 NASA's COBE (Cosmic Background Explorer) satellite team predicted and then discovered ripples in the cosmic background radiation. George Smoot, the team's leader, called these seeds for future galaxy superclusters, "fingerprints from the Maker." Dr. Smoot had a book published entitled *WRINKLES IN TIME*. This book is a report of his and the COBE satellite team's findings. A statement on the front cover of this book made by Stephen Hawking (who is reputed to be the world's greatest living mathematician and theoretical physicist), calls these findings, "The scientific discovery of the century, if not all time". The note on the outside of the back cover of the Avon Book's Trade Edition begins with a summary of the findings of this scientific team.

BEHOLD THE HANDWRITING OF GOD

"Astrophysicist and adventurer George Smoot spent twenty years pursuing the 'holy grail of science'—a relentless hunt that led him from the rain forests of Brazil to the frozen wastes of Antarctica. For decades he persevered—struggling against time, the elements, the forces of ignorance and bureaucratic insanity. And finally, on April 23, 1992, he made a startling announcement that would usher in a new scientific age. For George Smoot and his dedicated team of Berkeley researchers had proven the unprovable—uncovering, inarguably and for all time, the secrets of the creation of the Universe."

Dr. Smoot, with the help of over 1500 other scientists had demonstrated what is probably the most important discovery in the study of the creation of the universe since Galileo proved that Copernicus was correct when he stated that the earth was not the stationary center of the universe; that it rotated on its axis and orbited the sun. In *Wrinkles In Time,* pages 283–285, Dr. Smoot describes what happened when "God made heaven and earth". There are two things that must be remembered when reading Dr. Smoot's account of his discoveries. A light year is a measure of distance, not time; and evolution is used in its original meaning of "a change over time". It has nothing to do with the random chance theory of the origin of life.

"The evolution of the universe is effectively the change in distribution of matter over time moving from a virtual homogeneity in the early universe to a very lumpy universe today, with matter condensed as galaxies, clusters, superclusters, and even larger structures. We can view that change as a series of phase transitions, in which matter passes from one state to another under the influence of decreasing temperature . . . We are all familiar with the way that steam, on cooling, condenses. This is a phase transition from a gaseous to a liquid state. In the same way, matter has gone through a series of phase transitions since the first instant of the Big Bang."

"At a ten-millionth of a trillionth of a trillionth of a trillionth of a second after the big bang—the earliest moment about which we can sensibly talk, and then only with some suspension of disbelief—all the universe we can observe today was the tiniest fraction of the size of a proton. Space and time had only just begun. (Remember the universe did not expand into existing space after the big bang; its expansion created space, and time started at the moment the expansion began.) The temperature at this point was a hundred million trillion trillion degrees, and the three forces of nature—electromagnetism and the strong and weak nuclear forces—were fused as one. Matter was undifferentiated from energy, and particles did not yet exist."

"By a ten-billionth of a trillionth of a trillionth of a second inflation had expanded the universe (at an accelerating rate) a million trillion trillion times, and the temperature had fallen to below a billion billion billion degrees. The strong nuclear forces had separated, and matter underwent its first phase transition, existing now as quarks (the building blocks of protons and neutrons), electrons, and other fundamental particles."

"The next phase transition occurred at a ten-thousandth of a second, when quarks began to bind together to form protons and neutrons (and antiprotons and antineutrons). Annihilations of particles of matter and antimatter began, eventually leaving a slight residue of matter. All the forces of nature were now separate."

"The temperature had fallen sufficiently after the universe had expanded to a diameter of about a light minute, to allow protons and neutrons to stick together when they collided, forming the nuclei of hydrogen and helium, the stuff of stars. This soup of matter and radiation, which initially was the density of water, continued expanding and cooling until its diameter was three hundred thousand light years. At first, it was still too hot (energetic) for electrons to stick to the hydrogen and helium nuclei to form atoms. The energetic photons existed in a frenzy of interactions with the particles in the soup. The photons could travel only a very short distance between interactions. The universe was essentially opaque."[24]

"When the temperature fell to about 3,000 degrees, a crucial further phase transition occurred. The photons were no longer energetic enough to dislodge electrons from around hydrogen and helium nuclei and so atoms of hydrogen and helium formed and stayed together. The photons no longer interacted with the electrons and were free to escape and travel great distances."[24]

"With this decoupling of matter and radiation, the universe became transparent, and radiation streamed in all directions—to course through time as the cosmic background radiation we experience still. The radiation released at that instant gives us a snapshot of the distribution of matter when the universe had expanded to three hundred thousand light years. Had all matter been distributed evenly, the fabric of space would have been smooth, and the interaction of photons with particles would have been homogeneous, resulting in a completely uniform cosmic background radiation. Our discovery of the wrinkles reveals that matter was not uniformly distributed, that it was already structured, thus forming the seeds out of which today's complex universe has grown."[25]

"Cosmology, through the marriage of astrophysics and particle physics, is showing us that complexity flowed from a deep simplicity as matter metamorphosed through a series of phase transitions. Travel back in time through those phase transitions and we see an ever-greater simplicity and symmetry, with the fusion of the fundamental forces of nature and the transformation of particles to ever-more fundamental components. Go back further and we reach a point when the universe was nearly an infinitely tiny, infinitely dense concentration of energy, a fragment of primordial spacetime . . ."

"Go back further still, beyond the moment of creation, what then? What was there before the big bang? What was there before time began? Facing this, the ultimate question, challenges our faith in the power of science to find explanations of nature. The existence of a singularity, in this case the given, unique state from which the universe emerged, is anathema to science, because it is beyond explanation. There can be no answer to why such a state existed. Is this, then, where scientific explanation breaks down and God takes over, the artificer of that singularity, that initial simplicity?"

"The astrophysicist Robert Jastrow, in his book, *God and the Astronomers*, described such a prospect as the scientist's nightmare; "He has scaled the mountains of ignorance; he is about to conquer the highest peak; as he pulls himself over the final rock, he is greeted by a band of theologians who have been sitting there for centuries."[26]

"A useful analogy here is life itself, or, more simply, a single human being. Each of us is a vastly complex entity, assembled from many different tissues and capable of countless behaviors and thoughts. Trace that person back through his or her life, back beyond birth and finally to the moment of fertilization of a single ovum by a single sperm. The individual becomes ever simpler, ultimately encapsulated as information encoded in DNA in a set of chromosomes. The development that gradually transforms a DNA code into a mature individual is an unfolding, a complexification, as the information in the DNA is translated and manifested through many stages of life."[27]

So it was with the universe. The universe has metamorphosed from an infinitely small, infinitely dense, infinitely hot fused singularity, by a series of phase transitions, into the highly complex universe that we can now see, and human beings are only a very small part of that complexity.

Since space and time did not exist prior to creation, whoever made the infinitely small, dense, and hot singularity has to exist outside of space and time.

The discovery of wrinkles (phase transitions) in the expanding mass of matter when the universe had expanded to about 300,000 light years demonstrated that the matter of the universe was not uniformly distributed. It was already highly structured, forming the "seeds" from which our present, highly structured, universe has grown.

Two widely held beliefs among cosmologists, also known as cosmological principles, are:

1. The universe is homogeneous—the density of matter (galaxies) is the same everywhere.
2. The universe is isotropic; the distribution of matter (galaxies) is the same in every direction.

Olber's Paradox (Why is the night sky dark?) demonstrates that these beliefs are not true. If the universe is static, infinite, and uniform, then every line of sight must end on the surface of a star.

Why aren't we fried? The answers are, the universe had a beginning, the universe is expanding, the universe is finite, and the universe has not been uniform form the very beginning. In other words it is neither homogeneous nor isotropic.

The discovery of the wrinkles in time has demonstrated that these cosmological principles are not true. If these wrinkles had not occurred exactly when they did our present universe would not exist. For example "if the expansion rate of the universe at the beginning had been smaller by one part in one hundred thousand trillion, the universe would have recollapsed long ago. Expansion more rapid by one part in one million and there would have been no stars or planets."[28] In simple terms, had there been any variation in the timing of the development of the universe, it and we would not exist!

Where does this leave us with regard to the ultimate question? Did things just happen this way or is God so very smart that He got it just right the first time? It seems that the more we learn the more it all fits together. As we study the universe, we learn that the entire universe, from the smallest living cell to the great galactic masses, is not the consequence of a random series of meaningless events. It is the result of purpose and intelligent design.

"In 1977, Steven Weinberg published *The First Three Minutes*, one of the finest popular books on cosmology ever written. Toward the end of his book, Weinberg muses on the questions we ask ourselves, particularly the conviction that, somehow, humans are not a mere cosmic accident, the chance outcome of a concatenation of physical processes in a universe that dwarfs us on every scale. He expresses his view on the matter this way; 'It is very hard to realize that (this beautiful earth) is all just a tiny part of an overwhelmingly hostile universe. It is even harder to realize that this present universe has evolved from an unspeakably unfamiliar early condition, and faces a future extinction of endless cold or intolerable heat. The more the Universe seems comprehensible, the more it seems pointless."[29]

It is only pointless if there is no Creator or if He has not told us what His plan is. Actually the universe seems anything but pointless! "It seems that the more we learn, the more we see how it all fits together . . . how there is an underlying unity to the sea of matter and stars and galaxies that surround us . . . As we study the universe as a whole, we learn that nature is as it is not because it is the chance consequence of a random series of meaningless events; quite the opposite. More and more the universe appears to be as it is because it must be that way; its evolution was written in its beginnings in its cosmic DNA if you will. There is a clear order to the evolution of the universe . . . (Einstein had one right idea even if he didn't mean it as it sounds: 'God does not play dice with the universe'). The development of beings capable of questioning and understanding the universe seems quite natural."[30] If no such beings had been developed then the universe would truly be pointless!

"The religious concept of creation flows from a sense of wonder at the existence of the universe and our place in it. The scientific concept of creation encompasses no less a sense of wonder: We are awed by the ultimate simplicity and power of the creativity of physical nature and by it beauty on all scales."[31] You may call The Creator "the creativity of physical nature." You may not call Him random chance! We know Him to be God our Father.

Remember the words of Stephen Hawking; "The scientific discovery of the century, if not all time." Now let us compare Genesis 1:1–8, with Dr. Smoot's and the rest of the COBE scientists' findings.

Genesis 1:1–8: "**In the beginning God created heaven and earth. The earth was without form and void, with darkness over the face of the abyss, and the Spirit of God hovering over the sur-**

face of the waters. God said. 'Let there be light,' and there was light; and God saw that the light was good, and He separated light from darkness. He called the light day, and the darkness night. So evening came, and morning came, the first day."

"God said, 'Let there be a vault (empty space) between the waters, to separate water from water.' So God made the vault and separated the water under the vault from the water above it, and so it was; and God called the vault heaven. Evening came, and morning came, a second day."

If the Bible is Truth, and the Cosmic Background Radiation with its "Wrinkles in Time" (phase transitions) is really the "Handwriting of God", then the Biblical record in Genesis and the findings of the COBE scientists should be the same. The words may be different but the concepts should be the same.

The first sentence states that God made heaven and earth in the beginning, but the earth was without form and void; that is it had no shape, was empty, vacant, and contained no visible matter. If God made heaven and earth in the beginning, how could they be empty and not visible?

Now here is a paradox, in the real meaning of the word. A paradox is a statement that seems contradictory, unbelievable, or absurd but may actually be true. Until a few years ago, I did not really understand this paradox, and I did not believe it to be a factual statement. I thought it was a poetic description of the creation event, but now there is real, inarguable scientific evidence that this really occurred in the beginning.

Heaven and earth were created in the beginning. That is everything was there. Everything. This means the entire Universe was there. All of it was there, but it had no shape, it was vacant, there was nothing visible. It was pure force. The cosmologists of the COBE team, called this a singularity, and it was infinitely hot, infinitely dense, and infinitely small. If this seems hard to imagine, think of a fertilized human egg. All the information that is needed to become an adult human is there. All of that information is there in a structure smaller than this period (.) But it is all there.

Neither matter, nor energy, nor space existed, and time had not yet begun when God created heaven and earth. By the power of his WORD, God created a singularity that contained all matter and energy.

> John 1:2&3: "**The Word, then was with God at the beginning, and through Him all things came to be; no single thing was created without Him.**"

Thus the author of Genesis could write, "**The earth was without form and void**". Yet everything that was created was in that singularity!

Then God allowed the singularity to begin expanding. It did not explode. The expanding mass was shaped like a discus; it was rounded on the edges with a bulge in the center. Now remember, the universe did not expand into something when creation started. Space did not yet exist, and there were no clocks to record the beginning of time.

At first the universe was so compressed that it was pure force. As it expanded and cooled it went through a series of "wrinkles in time" or phase transitions. The first transition was force changing into things called quarks, electrons, and other fundamental particles. Apparently force can be changed into particles of matter by cooling. We are not sure how this happens, but we do know that it happened.

This first phase transition occurred when the expanded to one ten-billionth of a trillionth of light second (1.175×10^{-23} inches). The next change the universe had expanded to one ten thousandth of

(18.6 miles). The quarks began to form protons and neutrons, and all the forces that we know about in nature had separated.

When the universe had expanded to about one light minute (11,160,000 miles) the third change occurred and the nuclei of hydrogen and helium atoms began to form. The universe continued to expand and cool. When it had expanded to three hundred thousand light years the universe was still too hot and the particles were too close together for light to escape. It had the density of water, and all the particles were moving very rapidly due to the heat. They were packed so close to each other that they could not escape from the expanding mass. The first sentence in the first verse in Genesis was completed. (An abyss, according to the dictionary, is a great mass of water.)

> Genesis 1:1&2: "**In the beginning God created heaven and earth. The earth was without form and void, with darkness over the face of the abyss, and the Spirit of God hovering over the surface of the waters.**"

The Spirit of God was hovering over the face of the abyss until everything was exactly in place. When everything was exactly right, at that exact moment, The WORD spoke! "**Let there be light.**" The universe had cooled to about 3,000 degrees centigrade; hydrogen and helium atoms were forming; and light particles were able to escape from the expanding mass. These light particles (photons) did not escape randomly or in an even pattern. They traveled in the pattern that God had predetermined and for which the Spirit of God was watching and waiting.

> Genesis 1:4: "**And God saw that the light was good**".

Of course the light was good! If it had not been exactly right e Universe would have already collapsed or it would have formed ¡e cloud of hydrogen gas.

When the clumps of photons began escaping from the original expanding mass they still had the density of water. Then the Bible contains a statement indicating a major phase transition in the construction of the universe.

> Genesis 1:6–8: "God said, 'Let there be a vault (space) between the waters, to separate water from water.' So God made the vault and separated the water under the vault from the water above it, and so it was; and God called the vault heaven. (We call it space.) Evening came, and morning came, a second day."

At that exact moment photons were able to escape from the expanding mass and streamed, in clumps, creating the solar systems, the galaxies, in fact the entire universe. These clumps of photons were not equal in size or evenly distributed. They were the "seeds" from which all the galaxies, stars, and planets have grown.

In the beginning God created the universe, but He shaped it into an infinitely hot, infinitely dense, and infinitely small, singularity. When God allowed this singularity to expand, it became all of the structures of the universe, from the smallest atomic particle to the great galaxies. This did not just happen by chance. The entire universe was highly structured from the very beginning. "And God saw that it was good", exactly what and where He wanted it to be.

There is one sentence in Genesis that has caused some confusion among Christians.

> Genesis 1:5: He (God) called the light day, and the darkness night. I believe that John has given us the answer to this problem.

> 1 John 1:5: Here is the message we heard from Him and pass on to you: that God is light, and in Him there is no darkness at all.

From the very beginning we are to understand that where God is it is "day". Where God is not it is "night".

CHAPTER 12

Third From The Sun

A major phase transition occurred when God "divided" the original expanding matter (which had the density of water) into groups of photons. Another major phase transition occurred when God created the "vault" into which these groups of photons expanded. The author of Genesis tells us about this creation of the entire universe, including our solar system, in these words.

> Genesis 1: 6–8: "God said, 'Let there be a vault (space) between the waters, to separate water from water.' So God made the vault, and separated the water under the vault from the water above it, and so it was; and God called the vault heaven. Evening came and morning came, a second day."

The vault is part of the creation and is the area between the galaxies, stars, and planets. It is not totally empty but has so little matter in it that it was almost a pure vacuum. It is not eternity where God exists. The "waters" were the groups of photons that contained all the matter of the galaxies, but God had not yet cooled them enough to form the stars, planets, and all the other celestial bodies.

I was taught from childhood that the separation of the waters on the second day of creation represented the formation of lakes, rivers and oceans on earth and rain clouds in the sky. The major problem with that idea is, the stars, planets, and other celestial bodies had not yet been formed.

The author of Genesis gives no details of how the galaxies or solar systems were formed, and science has only theories. The only scientific evidence comes from our solar system itself, and all of the evidence that we have discovered so far, indicates that the earth was uniquely created to sustain life, as we know it.

A great deal of astronomical study has been done on the stars close enough to our solar system to be studied. When compared to our sun, which is just right to sustain life, the other stars are too big, too small, too hot, too cold, or are not composed of the proper elements in the right proportions. Only a few have planets orbiting them, and these planets are too big, too small, or are not composed of the proper elements in the right proportions. They also have highly elliptical orbits and are traveling at very high speeds. They also are very likely to be large and composed of mostly gas with no hard surface. Because of their elliptical orbits, their surface temperatures vary from near absolute zero to thousands of degrees centigrade. Because of their orbital speed their planetary year may last less than one earth week. There is no possibility that they could sustain any type of life, as we know it.

Steven Weinberg would have been right, except for the Earth and the Creator. "The more the Universe seems comprehensible, the more it seems pointless."[31] However, God had a plan!

Genesis 1:9&10: "**God said, 'Let the waters under heaven be gathered into one place so that dry land may appear;' and so it** was. God called the dry land earth, and the gathering of the waters He called seas; and God saw that it was good."

Science has only theories but no imperical evidence about how the earth might have been formed. The Cosmologists only know that the stars and planets were formed and that the fusion reaction began in the stars. Not how, why, or when this happened; only that it happened. We do not know, either, how the solar systems and galaxies were formed. Don Johanson, Director of the Institute of Human Origins at Arizona State University, made the following comment about Peter D. Ward and Donald Brownlee's book *Rare Earth.* "In this engaging and superbly written book, the authors present a carefully reasoned and scientifically astute examination of the age-old question—'Are we alone in the Universe?' Their astonishing conclusion that even simple animal life is most likely extremely rare in the Universe has many profound implications. To the average person, staring up at a dark night sky, full of distant galaxies, it is simply inconceivable that we are alone. Yet in spite of our wishful thinking, there just may not be other Mozarts or Monets." However the Bible does tell us what God's original plan for man was.

Genesis 1:26–31: "Then God said, 'Let Us make man in Our image and likeness to rule the fish in the sea, the birds of heaven, the cattle, all wild animals on earth, and all reptiles that crawl upon the earth.' So God created man in His own image; in the image of God He created him; male and female He created them. God blessed them and said to them. 'Be fruitful and increase, fill the earth and subdue it, rule over the fish in the sea, the birds of heaven, and every living thing that moves upon the earth.' God also said. 'I give you all plants that bear seed everywhere on earth, and ever tree bearing fruit which yields seed: they shall be yours for food. All green plants

I give for food to the wild animals, to all the birds of heaven, and to all reptiles on earth, every living creature.' So it was; and God saw all that He had made, and it was very good."

CHAPTER 13

Origin Of Life Fairy Tales

In the beginning, before the earth was solid, and while the solar system was being developed, nuclear reactions occurred that formed hydrogen atoms. Then more nuclear reactions occurred which converted these hydrogen atoms into all the other elements. This activity lasted about 400 million years. By the end of this time a solid planet earth was formed. This period is termed "Atomic Evolution".

Once the elements had formed they began to combine chemically into compounds that were more or less complex. Some carbon atoms combined with other elements to form organic compounds, and thus the "prebiotic (before life) organic chemical soup" was formed. This "organic chemical soup" was dispersed homogeneously through out the oceans of the earth. The atmosphere above these oceans contained many positively charged ions, but no free oxygen. Thus a reducing atmosphere was formed. This reducing atmosphere was necessary for amino acids and proteins to form.

Over a long period of time, six specific compounds combined in a very specific sequence to form a very large and specific molecule called deoxyribonucleic acid (DNA). In spite of being exposed to energy and the "soup" that should have dissolved this molecule, it was able to hold together. In fact a number of these molecules formed, did not dissolve, and learned how to take energy from the "soup" and reproduce themselves. This activity lasted about 1.5 billion years. This second period is termed "Chemical Evolution".

Over the next billion or so years, as the organic compounds in the "soup" were used up, these molecules changed into living cells. Some of these cells developed chlorophyll and learned how to change carbon dioxide into free oxygen. They also developed a nucleus and the ability to manufacture their own food from simple chemicals. During the next several billion years these simple cells changed into all of the millions of different kinds of living organisms that are known to be alive on the earth now, plus any other living things that have since become extinct, or have not yet been discovered.

When humans appeared and had risen above the animal level, the third period, "Organic Evolution" ended.

When humans had learned how to speak and write, they organized themselves into communities and this was the beginning of the fourth period. This period is still on going and is termed, "Cultural Evolution".

A Statement Affirming Evolution as a Principle of Science

For many years it has been well established scientifically that all known forms of life, including human beings, have developed by a lengthy process of evolution. It is also verifiable today that very primitive forms of life, ancestral to all living forms, came into being thousands of millions of years ago. They constituted the trunk of a "tree of life" that, in growing, branched more and more; that is, some of the later descendants of these earliest living things, in growing more complex, became ever more diverse and increasingly different from one another. Humans and the other highly organized types of today constitute the present twig-end of that tree. The human twig and that of the apes sprang from the same apelike progenitor branch.

Scientists consider that none of their principles, no matter how seemingly firmly established—and no ordinary "facts" of direct observation either—are absolute certainties. Some possibility of human error, even if very slight, always exists. Scientists welcome the challenge of further testing of any view whatever. They use such terms as firmly established only for conclusions founded on rigorous evidence that have continued to withstand searching criticism.

The principle of biological evolution, as just stated, meets these criteria exceptionally well. It rests upon a multitude of discoveries of very different kinds that concur and complement one another. Scientists and other reasonable persons who have familiarized themselves with the evidence therefore accept it into humanity's general body of knowledge.

In recent years, the evidence for the principle of evolution has continued to accumulate. This has resulted in a firm understanding of biological evolution, including the further confirmation of the principle of natural selection and adaptation that Darwin and Wallace over a century ago showed to be an essential part of the process of biological evolution.

There are no alternative theories to the principle of evolution, with its "tree of life" pattern, that any competent biologist of today takes seriously. Moreover, the principle is so important for an understanding of the world we live in and of ourselves that the public in general, including students taking biology in school, should be made aware of it, and of the fact that it is firmly established in the view of the modern scientific community.

Creationism is not scientific; it is a purely religious view held by some religious sects and persons and strongly opposed by other religious sects and persons. Evolution is the only presently known strictly scientific and nonreligious explanation for the existence and diversity of living organisms. It is therefore the only view that should be expounded in public school courses on science, which are distinct from those on religion.

We, the undersigned, call upon all local school boards, manufacturers of textbooks and teaching materials, elementary and secondary teachers of biological science, concerned citizens, and educational agencies to do the following:

Resist and oppose measures currently before several state legislatures that would require creationist views of origins are given equal treatment and emphasis in public school biology classes and text materials.

Reject the concept, currently being put forth by certain religious and creationist pressure groups, that alleges that evolution is itself a tenet of a religion of "secular Humanism," and as such is unsuitable for inclusion in the public school science curriculum.

Give vigorous support and aid to those classroom teachers who present the subject matter of evolution fairly and who often encounter community opposition."

More than 500 scientists and college professors signed this Affirmation.

These attempts to theorize how life began are interesting fables. The paraphrase is from a college textbook *The Evolutionary Process*[32] used in Universities as recently as 1993. This book is no longer available from the publisher, but the story is still being taught in the public schools as if it were TRUTH. The statement affirming evolution as a principle of science first appeared in the *Humanist* of January/February, 1977. There has been no retraction of this affirmation.

The statement that animal life has evolved by random chance, from very simple to very complex organisms, over millions of years, is simply a lie or an example of gross ignorance. The Burgess Shale fossil site was discovered in 1909 but was not reported to the public until the late 1980's when the Chengjiang fossil site was discovered in China. Both of these sites show that every phyla of living animal that is present on earth now and at least 12 other phyla that are not present now, were present in the "very, very beginning", and no intermediate types were found in either site. Yet the "tree of life" theory is still being taught in our schools and universities as if it were true. The next chapter illustrates how this can happen.

Chapter 14

The Emperor's New Clothes

Many years ago there was an Emperor who was so very fond of new clothes that he spent all his money on dress. He did not trouble himself in the least about his soldiers, nor did he care to go either to the theater or to hunt, except for the occasion they gave him for showing off his new clothes. He had a new suit for each hour of the day; and as of any other king or emperor one is accustomed to say, "He is sitting in council," it was always said of him, "The Emperor is sitting in his wardrobe."

Time passed merrily in the large town that was his capital. Strangers arrived at the court every day. One day two rogues, calling themselves weavers, made their appearance. They gave out that they knew how to weave stuffs of the most beautiful colors and patterns, but that the clothes made from these had the wonderful property of remaining invisible to every one who was either stupid or unfit for the office he held.

"Those would indeed be splendid clothes!" thought the Emperor. "Had I such a suit, I might at once find out what men in my realms are unfit for their office, and be able to distinguish the wise from the foolish. This stuff must be woven for me immediately."

And he caused large sums of money to be given to the weavers, that they might begin their work at once.

So the rogues set upon two looms, and made a show of working very busily, though in reality they had nothing at all on the looms. They asked for the finest silk and the purest gold thread, put both into their own knapsacks, and then continued their pretended work at the empty looms until late at night.

"I should like to know how the weavers are getting on with my cloth", thought the Emperor after some time. He was, however, rather nervous when he remembered that a stupid person, or one unfit for his office, would be unable to see the stuff. "To be sure," he thought. "I have nothing to risk in my own person but yet I would prefer sending somebody else to bring me news about the weavers and their work, before I trouble myself in the affair." All the people of the city had heard of the wonderful property the cloth was to possess, and all were anxious to learn how worthless and stupid their neighbors were.

"I will send my faithful old minister to the weavers," concluded the Emperor at last. "He will be able to see how the cloth looks, for he is a man of sense, and no one can be better fitted for his post than he is."

So the faithful old minister went into the hall where the knaves were working with all their might at their empty looms. "What can be the meaning of this?" thought the old man, opening his eyes very wide. "I can't see the least bit of thread on the looms!" However, he did not speak aloud.

The rogues begged him most respectfully to be so good as to come nearer, and then asked whether the design pleased him and whether the colors were not very beautiful, pointing at the same time to the empty frames. The poor old minister looked and looked; he could see nothing on the looms, for there was nothing there. "What!" thought he "Is it possible that I am silly? I have never thought so myself, and no one must know it now. Can it be that I am unfit for my office? It will never do for me to say that I could not see the stuff."

"Well, Sir Minister!" said one of the knaves, still pretending to work. "You do not say whether the stuff pleases you."

"Oh, its very fine!" said the old minister, looking at the loom through his spectacles. "The pattern and the colors are wonderful. Yes, I will tell the Emperor without delay how very beautiful I think them."

"We are glad they please you," said the cheats, and then they named the different colors and described the pattern of the pretended stuff. The old minister paid close attention that he might repeat to the Emperor what they said.

Then the knaves asked for more silk and gold, saying it was needed to complete what they had begun. Of course, they put all that was given them into their knapsacks, and kept on as before working busily at their empty looms.

The Emperor now sent another officer of his court to see how the men were getting on, and to find out whether the cloth would soon be ready. It was as just the same with him as with the first. He looked and looked, but could see nothing at all but the empty looms.

"Isn't it fine stuff?" asked the rogues. The minister said he thought it beautiful. Then they began as before, pointing out its beauties and talking of patterns and colors that were not there.

"I certainly am not stupid," thought the officer. "It must be that I am not fit for my post. That seems absurd. However, no one shall know it." So he praised the stuff he could not see, and said he was delighted with both colors and patterns. "Indeed, Your Majesty," said he to the Emperor when he gave his report, "the cloth is magnificent."

The whole city was talking of the splendid cloth that the Emperor was having woven at his own cost.

And now the Emperor thought he would like to see the cloth while it was still on the loom. Accompanied by a select number of officials, among whom were the two honest men who had already admired the cloth, he went to the cunning weavers who, when aware of the Emperor's approach, went on working more busily than ever, although they did not pass a single thread through the looms.

"Is it not absolutely magnificent?" said the two officers who had been there before. "If Your Majesty will only be pleased to look at it! What a splendid design! What glorious colors!" And at the same time they pointed to the empty looms, for they thought that everyone else could see the cloth.

"How is this?" the Emperor said to himself, "I can see nothing! Oh, this is dreadful! Am I a fool? Am I unfit to be an Emperor? That would he the worst thing that could happen to me". Oh! The cloth is charming," said he aloud. "It has my complete approval." And he smiled most graciously, and looked closely at the empty looms; for on no account would he say that he could not see what two of the officers of his court had praised so much. All the retinue looked and looked, but they could see nothing anymore than the others. Nevertheless, they all exclaimed. "Oh how beautiful!" and advised His Majesty to have some new clothes made from this splendid material for the approaching procession. "Magnificent! Charming! Excellent!" resounded on all sides; and everyone seemed greatly pleased. The Emperor showed his satisfaction by making the rogues knights, and giving them the title of "Gentlemen Weavers to the Emperor".

The two rogues sat up the whole of the night before the day of the procession. They had sixteen candles burning, so that everyone might see how hard they were working to finish the Emperor's new suit. They pretended to roll the cloth off the looms; they cut the air with great scissors, and sewed with needles without any thread in them. "See!" cried they at last; "the Emperor's new clothes are ready!"

And now the Emperor, with all the grandees of his court, came to the weavers. The rogues raised their arms, as if holding something up, and said, "Here are Your Majesty's trousers! Here is the scarf! Here is the mantle! The whole suit is as light as a cobweb. You might fancy that you had on nothing at all when dressed in it; that, however, is the great virtue of this fine cloth."

"Yes, indeed!" said all the courtiers, although not one of them could see anything, because there was nothing to be seen.

"If Your Imperial Majesty will be graciously pleased to take off your clothes, we will fit on the new suit in front of the large looking glass," said the swindlers.

The Emperor accordingly took off his clothes, and the rogues pretended to put on him separately each article of his new suit, the Emperor turning round from side to side before the looking glass.

"How splendid His Majesty looks in his new clothes! And how well they fit!" everyone cried out. "What a design! What colors! These are indeed royal robes!"

"The attendants are waiting outside with the canopy which is to be borne over Your Majesty in the procession." announced the chief master of the ceremonies.

"I am quite ready," announced the Emperor. "Do my new clothes fit well?" he asked. He turned himself round again before the looking glass as if he were carefully examining his handsome suit.

The lords of the bedchamber, who were to carry His Majesty's train, felt about on the ground, as if they were lifting up the ends of the mantle, and walked as if they were holding up a train; for they feared to show that they saw nothing and so be thought stupid or unfit for their office.

So in the midst of the procession the Emperor walked under his high canopy through the streets of his capital. And all the people standing by, and those at the window, cried out, "Oh! How beautiful are our Emperor's new clothes! What a train there is to the mantle! And how gracefully the scarf hangs!" In short, no one would allow that he could not see those much admired clothes, because, in doing so, he would have declared himself either a fool or unfit for his office. Certainly, none of the Emperor's previous suits had made such an impression as this.

"But the Emperor has nothing on at all!" said a little child.

"Listen to the voice of innocence!" exclaimed her father, and what the child had said was whispered from one to another.

"But he has on nothing on at all!" at last cried out all the people.

The Emperor was vexed, for he felt that the people were right; but he thought the procession must go on now. And the lords of the bedchamber took greater pains than ever to appear to be holding up a train, although, in reality, there was no train to hold.[33]

Adolph Hitler, in *Mein Kampf* ch.10 wrote; "The great masses of the people will more easily fall victims to a great lie than to a small one."

Revelation 3:17: "You say, 'how rich I am! And how well I have done! I have everything I want.' In fact, thought you do not know it, you are the most pitiful wretch, poor, blind, and naked."

To know the truth we would do well to be as observant as the child in the fairy tale.

Isaiah 11:6: ". . . And a little child shall lead them."

CHAPTER 15

Life By Abiogenesis

We have irrefutable scientific evidence that the universe was created by the expansion of an infinitely small, infinitely dense, infinitely hot, but highly structured singularity. There is also irrefutable evidence that this expansion occurred in a timed sequence that was accurate to trillionths of a second. The next question is: was life created the same way or did it just happen by random chance?

Lets agree on the meaning of a few more words before we go any further.

1. Creation is a statement of belief that an immortal, intelligent, personal, being who is not part of the creation has created the universe and everything in it. In this context the term God will be used to name this Being.
2. Time is the period from the creation to the end of the universe.
3. Space is the area in eternity that is occupied by the universe.
4. Evolution is used in its original meaning of "change over time", but no outside force had anything to do with the

change. Natural processes alone operated to form life on this planet. No mysterious, divine, or vital (living) forces had a part in any of the changes.

5. The terms Creation and Evolution are both statements of belief, and are about origins; one-time events that were not observed. Holders of both beliefs agree that the universe is real and can be studied by using the scientific method, but the singular i.e. one time events of both creation and evolution were not observed, and thus cannot be tested.

6. The <u>scientific method</u> is inductive and involves the formation of theories that can be tested by experimentation and observation.

7. A <u>fact</u> is something that is true and accurate, a thing that has real, demonstrable existence. In science a fact is something that can be demonstrated by repeated experimentation.

Why all the stress on testing theories? The scientific process is inductive and requires formation of postulates about things that have been observed; theories about how these things happened; and experiments that can be observed and repeated that will verify or falsify the theory and the postulates. Since "the creation of living things" and "the evolutionary development of living things by random chance", are postulates about origins and were not observed, theories and experiments about them can only be studied by statistical analysis and can only show probability. Every time an experiment about an origin postulate fails, it decreases the probability of that postulate being true, and increases the probability that some other postulate is true.

Lets look at evolution first. Remember that for a theory to be studied scientifically it must be subjected to experimentation and observation. Drs. Thaxton, Bradley, and Olsen, in their textbook *The Mystery of Life's Origin,* have pointed out that there are only two ways that biochemical evolution (the formation of living matter from non-living matter) is possible. These are <u>spontaneous generation</u>, the spontaneous formation of living matter from inorganic matter and <u>heterogenesis</u>, spontaneous formation of living matter from dead organic matter. Organic matter is made of carbon containing compounds. All other kinds of matter are inorganic.

Spontaneous generation was abandoned in 1864. Pasteur showed that air contains many microorganisms that can collect and multiply in water, giving the illusion of spontaneous generation. Pasteur announced his results before the science faculty at the Sorbonne in Paris with the words; "Never will the doctrine of spontaneous generation recover from the mortal blow of this simple experiment."[34] Heterogenesis was abandoned because it could not explain the origin of the dead organic matter.

Darwin himself and many others recognized that his work now required an even more difficult and remarkable form of spontaneous generation, <u>abiogenesis</u>. In 1871 Darwin wrote in a letter: "It is often said that all the conditions for the first production of a living organism are now present which could ever have been present. But if (and oh! what a big if!) we could conceive in some warm little pond, with all sorts of ammonia and phosphoric salts, light, heat, electricity, etc. present, that a protein compound was chemically formed ready to undergo still more complex changes, at the present day such matter would be instantly devoured or absorbed, which would not have been the case before living creatures were formed."[35]

In the sixth edition of *The Origin Of Species*, published in 1872 the last paragraph of the book has this to say.

"It is interesting to contemplate a tangled bank, clothed with many plants of many kinds, with birds singing on the bushes, with various insects flitting about, and with worms crawling through the damp earth, and to reflect that these elaborately constructed forms, so different from each other, and dependent upon each other in so complex a manner, have all been produced by laws acting around us. These laws, taken in the largest sense, being Growth with reproduction; Inheritance which is almost implied by reproduction; Variability from the indirect and direct action of the conditions of life, and from use and disuse; a Ratio of Increase so high as to lead to a struggle for life, and as a consequence of Natural Selection, entailing Divergence of Character and the Extinction of less-improved forms. Thus, from the war of nature, from famine and death, the most exalted object which we are capable of conceiving, namely, the production of the higher animals, directly follows. **There is grandeur in this view of life, with its several powers, having been originally been breathed by the Creator into a few forms or into one**; and that, whilst this planet has gone cycling on according to the fixed law of gravity, from so simple a beginning endless forms most beautiful and most wonderful have been, and are being evolved."[36]

Darwin almost had it right! If he had known about the fossil sites in Canada and China, and the "Language of life" (DNA) it is very likely that the whole debate over *The Origin Of Species* would never have occurred.

Darwin's comments in 1871 were the first suggestion leading to the modern experiments attempting to prove that abiogenesis did occur. This was followed in 1924 when the Russian biochemist Alexander Ivanovich Oparin proposed that the complex molecular arrangements and functions of living systems could have evolved from simpler molecules that preexisted on the lifeless primitive earth.[37]

In 1928, British biologist J.B.S. Haldane suggested that ultraviolet light acting upon the earth's primitive atmosphere would result in an increasing concentration of sugars and amino acids in the ocean. He believed that life might eventually emerge from this "primordial soup".[38]

In 1947 J.D. Bernal suggested some possible mechanisms whereby biomonomers might accumulate in concentrations sufficient to allow condensation reactions that would produce the macromolecules necessary for life.[39]

In 1952 Harold Urey observed that with the exception of the earth and the minor planets, the rest of the planets of this solar system with atmospheres, had reducing atmospheres, a condition necessary for the synthesis of organic compounds. Perhaps the early earth's atmosphere had been reducing and only became oxidizing later in its evolution.[40]

These suggestions by Darwin, Oparin, Haldane, Bernal, and Urey became the foundation for what is known as the modern theory of chemical evolution, or Darwinism. This theory has continued to dominate the thinking of some scientists into the 1990's. It consists of five major stages.

The first stage envisions the early earth atmosphere as containing significant amounts of hydrogen, methane, carbon monoxide, carbon dioxide, ammonia, and nitrogen, but no free oxygen. This would provide the reducing atmosphere necessary for the formation of organic molecules. At first these molecules would have broken down as fast as they were formed, but when the earth's surface had cooled to less than 100°C some of the molecules would be able to survive.

During the second stage various forms of energy would have been available to drive the reactions that would form a wide variety of organic molecules. These energy sources would have included lightning, geothermal heat, shock waves, ultraviolet light from the sun, and others. Since there was no free oxygen to form an ozone layer, the ultraviolet light would have irradiated the reducing atmosphere to produce amino acids, formaldehyde, hydrogen cyanide and many other compounds.

Other compounds would have formed at lower altitudes. The energy needed for the formation of these compounds would have been supplied by electrical storms and thunder shock waves. Organic compounds would have formed on the surface as the atmospheric gases were blown over hot lava flows. All of these simple compounds would have been washed into the ocean by rain and wind. This mixing in the ocean would have produced further reactions, and more complex compounds. Eventually this would produce an ocean with the consistency of "hot dilute soup."

During the third stage this soup would have washed over the tidal flats leaving behind pools and lagoons where the water would evaporate and thicken the soup. Sinking clay particles would have then catalyzed these compounds to form polymeric macromolecules including peptides and polynucleotides.

Everything was now ready for stage four, the formation of protocells. These are not true living cells but they have a membrane and sufficient functional capacity to survive long enough to become true cells. Once the polypeptides had become enzymes, and other characteristics of living cells had emerged, nucleic acids would have formed and became sufficiently developed to form DNA and life would have begun. Stage five was now complete. <u>These are wonderful sounding words but there is no experimental evidence that any of this ever happened!</u>

Many experiments have been devised and performed in an attempt to demonstrate the truth of this theory, but before looking at these, let's consider how origin-of-life research relates to science as a whole.

"In the matter of origins, there were no observers present. For some this lack of observation entirely removes the question of life's origin from the domain of legitimate science."[41]

"It is inherent in any acceptable definition of science that statements that cannot be checked by observation are not really about anything, or at the very least they are not science."[42]

"It is this lack of observational check on theories, that makes it impossible for science to provide any empirical knowledge (facts based on experiment and observation) about origins. It can only suggest scenarios in an attempt to reconstruct the events that led to the appearance of life on earth.

"Those who work on the origin of life must necessarily make bricks without very much straw, which goes a long way to explain why this field of study is so often regarded with deep suspicion. Speculation is bound to be rife, and it has also frequently been wild. Some attempts to account for the origin of life on the Earth, however ingenious, have shared much with imaginative literature and little with theoretical inference of the kind which can be confronted with observational evidence of some kind or another."[43]

All of this does not mean that origin-of-life experiments should not be undertaken, only that they will not verify or falsify the basic belief. They will only demonstrate probability and not proven fact, and the odds are long indeed. "The probability that at ordinary temperatures a macroscopic number of molecules is assembled to give rise to the highly ordered structures and to the coordinated functions characterizing living organisms is vanishingly small. The idea of spontaneous generation of life in its present form is therefore highly improbable even on the scale of the billions of years during which prebiotic evolution occurred."[44]

Some scientists were, and still are, willing to bet very large sums of money that these odds can be beaten! "We are confident that the basic process (of chemical evolution) is correct, so confident that it seems inevitable that a similar process has taken place on many other planets in the solar system . . . We are sufficiently confident of our ideas about the origin of life that in 1976 a spacecraft will be sent to Mars, to land on the surface, with the primary purpose of the experiments being a search for living organisms."[45]

In 1976, on the eve of the first Mars landing NASA's chief biologist, Harold P. Klein, explained that, "If our theories of origins are correct, we should find corroborative evidence of it on Mars."[46]

Unfortunately neither the Mars landing, nor the studies of Venus, nor the Voyager I fly—by of Jupiter and Saturn, nor the studies of Titan (one of the moons of Saturn) have yielded any evidence of life, or the possibility of the origin of life.

The second Mars probe, launched in 1998, has found no evidence that life of any kind has ever existed on that planet. During 1999 a Mars Lander crashed on landing, and later that year a second Lander failed to communicate at all with Earth. There is still no evidence that any form of life as we know it has ever existed on Mars.

In summary, "Chemical evolution is broadly regarded as a highly plausible scenario for imagining how life on earth might have begun. It has received support from many competent theorists and experimentalists. Ideas of chemical evolution have been modified and refined considerably through their capable efforts. Many of the findings of these workers, however, have not supported the scenario of chemical evolution. In fact, what has emerged over the last several decades is an alternative scenario which is characterized by destruction, and not the synthesis of life."[47]

These findings have included a primitive earth with an oxidizing atmosphere where no chemical evolution could have occurred. If it existed at all, the prebiotic chemical soup, necessary for the formation of essential precursor chemicals, was so dilute that it left no known trace in the geological record. Finally no amount of energy flow through any random system has produced any chemical that contains any biological information.

Many well-designed experiments as well as some poorly designed experiments have been performed in an effort to prove that Darwinian abiogenesis could have occurred. Not one has been able to demonstrate that life could have originated under the conditions being studied. Since each failed experiment reduces the probability of abiogenesis having occurred, we may safely conclude that the formation of living organisms by the process of abiogenesis is highly unlikely.

"The problem is not what we **do not** know but what we **do** know. Many facts have come to light in the past few decades of experimental inquiry into life's beginnings. With each passing year the criticism has gotten stronger. The advance of science itself is what is challenging the notion that life arose on earth by spontaneous chemical reactions."[48]

"The idea of Darwinian molecular evolution is not based on science. There is no publication in the scientific literature (in journals or books) that describes how molecular evolution of any real, complex, biochemical system either did occur or even might have occurred. There are assertions that such evolution occurred, but absolutely none are supported by pertinent experiments or calculations. Since there is no authority on which to base claims of knowledge of such evolution, it can truly be said that the assertion of Darwinian molecular evolution is merely bluster."[49]

CHAPTER 16

The Origin Of Life

What observations, if any, support the belief that life was created by an intelligent, personal, being who is not part of the creation? Recent findings while studying deoxyribonucleic acid (DNA) have convinced many scientists that the structure of DNA is so complex that it is impossible for it to have formed by random chance. Furthermore it is not only the complexity of the structure, it is also the information contained in DNA itself. It has been said that for the information contained in DNA to occur by random chance would be similar to a tornado moving through a large junkyard and leaving a completely assembled and flyable 747 jet aircraft behind it. This is a reasonable simile, but a better one is the same tornado going through the same junkyard and leaving a complete set of blue prints, all the necessary parts, the machinery, and all the people necessary to assemble the 747.

In fact it is even more difficult than that. DNA is not simply a blue print; it is a complete language, that includes a decoder, a translator, and a messenger to carry the message to the workers who will use the information to build the specific parts of the living structure that they are building! When the president of the

United States announced the completion of the Human Genome project, he called the complete sequence of human DNA, the "Book of Life".

Let us consider some aspects of language to enable us to understand this "Book of Life". All languages can be reduced to writing by using a variety of symbols or representations. In the English language we call these symbols, letters. There are 26 letters in the English alphabet and one or a combination of these 26 letters can represent each word in the English language. DNA is also a language, and it is made up of six chemical compounds that are the foundations for the four letters in the DNA alphabet. These compounds are, a carbohydrate (deoxyribose), four bases (adenine, cytosine, guanine, and thymine), and a phosphate group. Biochemical compounds are optically rotated either to the left (left handed) or to the right (right handed). A random mixture of a biochemical compound will contain equal amounts of left handed and right handed forms. This is called a racemic mixture

Each letter in the DNA language is made up of the carbohydrate, one of the bases, and the phosphate group. In order to carry any information these compounds must be combined very specifically. Each DNA letter contains one right-handed carbohydrate, one left handed base, and one right handed phosphate group. Each DNA letter is called a nucleotide and can be combined with other DNA nucleotides in specific arrays. In order to carry any useful information, these arrays must be connected in a specific order.

The carbohydrate must be connected to the phosphate group and to one of the bases as a side chain. This arrangement allows a long chain to form that must be connected carbohydrate to phosphate to carbohydrate to phosphate—etc—etc—etc, with a base as a side chain connected to each carbohydrate. Two long strands are formed with one going c—p—c—p . . . left to right and the other going c—p—c—p . . . right to left. The long chains are twisted around each other with the bases facing inward. They are arranged so that adenine is always opposite thymine and cytosine is always opposite guanine. Each pair of adenine and thymine bases is connected by a double hydrogen bond. Each pair of cytosine and guanine bases is connected by a triple hydrogen bond. These bonds hold the entire molecule tightly together in an array called a double helix.

For a language to carry any useful information it must be both specific and complex. The phrase (I You Me) repeated over and over indefinitely is specific but carries very little useful information. However a book such as the Bible, using only the 26 letters of the English language, is both specific and complex. That is, the letters are specific and the arrangement of the letters is complex so that they carry a great deal of information for the one who can read English. DNA is equivalent to English or any other readable language in that it is both specific and complex.

The DNA molecule contains all of the information necessary for the germ cell of any particular species to grow from a single cell to an adult individual. In addition it carries the information needed to produce new cells each with a complete copy of its own DNA. There is also a program to proof read the copy and correct mistakes, and enough redundancy to compensate for any errors missed by the proofreader.

The following figure is a schematic diagram of a possible part of a DNA molecule.[50]

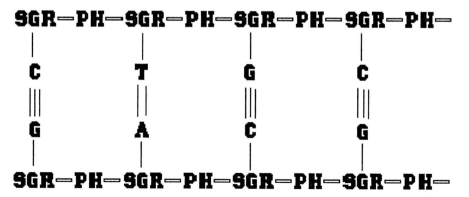

SGR is Deoxyribose; PH is the phosphate group; C, T, G, and A, are the bases cytosine, thymine, guanine, and adenine. Each group of deoxyribose, phosphate, and a base is one letter in the DNA alphabet. These letters are called nucleotides and there are only four of them in the DNA alphabet. However each nucleotide is made up of three compounds and each compound has a number of atoms. Deoxyribose is a carbohydrate with 20 atoms, five carbon, five oxygen, and ten hydrogen. The phosphate group has 5 atoms, one phosphorous and four oxygen. The four bases contain from 12 to 16 atoms each. Cytosine has 13 atoms, four carbon, five hydrogen, three nitrogen, and one oxygen. Thymine has 15 atoms, five carbon, six hydrogen, two nitrogen, and two oxygen. Guanine has 16 atoms, five carbon, five hydrogen, five nitrogen, and one oxygen. Adenine has 15 atoms, five carbon, five hydrogen, and five nitrogen.

Using simple probability there are (20x5x13) + (20x5x15) + (20x5x16) + (20x5x15) or 6225 possible arrangements of these atoms in the four nucleotides, but only 4 specific arrangements are valid letters in the DNA alphabet. If you assume that these four specific arrangements of atoms did occur by chance alone, then you are faced with the remote probability that these four compounds reproduced themselves and formed, by chance alone, the complex structure of DNA.

There are approximately 3 billion nucleotides in the average mammalian DNA molecule, with many possible sequences of these nucleotides. The actual number is impossibly large. There are in fact about $10^{24,082,400}$ possible sequences. This is the number one followed by more that twenty four million zeros. In addition some flowering plants have over 300 billion nucleotides in their DNA sequence, which makes the number of possible sequences for flowering plants about $10^{2,408,240,000}$. The number for plants is one followed by two billion four hundred million zeros, and plants had to come first! Green plants replenish the oxygen in the Earth's atmosphere. Without these plants, life could not exist on Earth!

With the completed sequencing of the human genome in 1999, we now know the order of the letters in the human "Book of Life". We are only just beginning to learn the words, sentences, and paragraphs in this book. It is obviously a written language and the book is species specific. There is no possibility of a book of this complexity and specificity happening by random chance. There are some Darwinists who admit that the probability of random chance producing DNA is impossibly remote, but they still maintain that it had to happen because they do not believe that a CREATOR exists.

In mid February 1999, the narrator of a television broadcast on the "Discovery Channel", while discussing the origin of life on earth, stated that the idea of a "prebiotic organic chemical soup" as the origin of living organisms had been abandoned. It has been recognized that the formation of DNA by random chance simply cannot happen. So now there is a new theory.

Since the surface of Mars appears to be pitted by many collisions with comets and asteroids, it was theorized that some of these comets or asteroids contained "spores" that were alive but in suspended animation. When the collisions with Mars broke up the comets and asteroids, some of the pieces containing these "spores" drifted away from Mars.

Millions of years later they drifted to Earth. Because the conditions on Earth were suitable, the spores germinated and began to reproduce. After several millions of years and enough micro-mutations these spores evolved into all of the life forms found on Earth today. However this theory does not explain the origin of the spores!

This theory was nullified a few days later, on the same television program, when it was announced that the pitting on Mars was not from comet and asteroid collisions. The pits were actually cones formed by volcanic activity from the interior of the planet. While spores might survive the cold of space, they could not have survived the heat produced by the volcanic activity.

The on going "Search for Extraterrestrial Intelligence" has so far found a few stars with planets but no planetary system that has any possibility of forming or sustaining life. The only evidence that we have about the formation of life is the written message that is present in the DNA of every living cell of every living organism on Earth, both plants and animals, each with seed according to its kind.

Now let us find out what the Creator had to say about how life began on earth. Genesis 1 begins **"In the beginning God created"** and proceeds to outline in poetic form how God created matter, then the universe, and then the earth. Then He prepared the earth for living creatures.

> Genesis 1:11: **"Then God said. 'Let the earth produce fresh growth, let there be on the earth plants bearing seed, fruit-trees bearing fruit each with seed according to its kind'."**

We really need that oxygen! Then God started the fusion reaction in the sun and the rest of the stars.

> Genesis 1:14: **"God said, 'Let there be lights in the vault of heaven to separate day from night, and let them serve as signs both for festivals and for seasons and years'."**

This explains why there has been such confusion about time. God's time is recorded in days and festivals and seasons and years. The time that man has invented, trying to be more accurate than God, is recorded in seconds and minutes and hours. God's time is based on the motion of the universe. Man's time is based on the motion of mechanical devices called clocks that are made by man. Man's devices can be very regular but they do not correspond with the motion of God's creation.

> Genesis 1:20&21: "Then God said, 'Let the waters teem with countless living creatures, and let birds fly above the earth' . . . God then created the great sea-monsters and all living creatures that move and swarm in the waters, according to their kind, and every kind of bird".

> Genesis 1:24: "Then God said, 'Let the earth bring forth living creatures, according to their kind'."

> Genesis 1:26–31: "Then God said, 'Let US make man in OUR image and likeness' . . . So God created man in His own image; in the image of God He created him; male and female He created them' . . . and God saw all that He had made, and it was VERY GOOD."

Wasn't it thoughtful of God to give us a handbook that not only tells us what He did, but also tells us where to find the proof of how He did it? "Each with seed according to its kind." It has only taken us a few thousand years to rediscover that God put a written message in each living thing to tell us that He is the Creator. Incidently the only thing that God created after the beginning was man! All the chemicals were there but God did not put them together until the earth was ready to sustain human life.

What about the account in Genesis 2 and the rest of the Bible? There is an apparent difference between chapter one and chapter two in the sequence of things created. Is this a real difference or can it be explained logically? Since the Bible is silent about this difference any theory of why it exists is only a theory, it is not essential to knowing God! However I like to speculate, so this is my own theory. It may or may not be true; only time will tell.

The first verse of the first chapter of Genesis tells us that the heavens and the earth were created in the beginning. Since the statement is in the past tense it implies a completed event. Satan knew that God would create a perfect universe and he wanted to spoil it right from the beginning. The first creation account lays out God's perfect plan, but something happened. Satan began tinkering with creation from the very beginning. Starting at Genesis 2:5 we have the account of what happened because Satan was trying to destroy God's universe, but that is another story.

CHAPTER 17

The End Of The Age

Some people believe that the Universe will last forever and others believe that, just as it had a beginning, it will have an end. Think back for a few minutes to the statement by Carl Sagan; "The Cosmos is all that is, or ever was, or ever will be." The major reason for this book is to show that this Humanist worldview of is not only wrong, but it is easy to prove scientifically that it is wrong. If Sagan was right then the Cosmos (Universe) had no beginning and will never end. But we have proven that the Universe had a beginning!

Dr. Smoot and the other scientists working on the COBE project spent twenty years pursuing this "holy grail of science". On April 23, 1992, they made the announcement that would usher in a new scientific age. For George Smoot and his dedicated team of Berkeley researchers had proven the unprovable, uncovering inarguably and for all time, the secrets of the creation of the Universe. The discovery of the "background radiation" and the phase transitions during the initial expansion of the Universe, have proven that there was a beginning.

What proof is there that the Universe will end? The structure of the Universe itself gives us all the proof needed to know that the Universe will have an end.

When the first fission bomb was dropped on Hiroshima we had our first inkling of what real power is. Then, with the world's first thermonuclear detonation at Eniwetok Proving Grounds on November 1, 1952, we had a slightly larger demonstration of the power of the force (shock wave) that is released by our sun and all the other active stars in the Universe, when hydrogen is fused into helium.

But remember, the explosion at Eniwetok was a one-time event that only lasted a fraction of a second, whereas the fusion reaction in the stars is a series of very close together thermonuclear detonations. The force or shock waves from these detonations is not just a moving wall of force that goes by and then is gone. It is a series of rapid bumps of varying strengths that keeps recurring as long as there is any hydrogen left in the star.

But when the hydrogen is used up, the fusion reaction stops, the shock wave stops being produced, and gravity takes over. Then the universe will collapse into a black hole.

As Steven Weinberg put it in his book *The First Three Minutes*, "It is very hard to realize that (this beautiful earth) is all just a tiny part of an overwhelmingly hostile universe. It is even harder to realize that this present universe has evolved from an unspeakably unfamiliar early condition, and faces a future extinction of endless cold or intolerable heat." [30]

Christianity teaches that Creation occurred "in the beginning". What does it teach about the end?

Matthew 24:3–36: "**When Jesus was sitting on the Mount of Olives the disciples came to speak to him privately. 'Tell us,' they said, 'when will this happen? And what will be the signal for your coming and the end of the age?'**

"Jesus replied: 'Take care that no one misleads you, for many will come claiming my name and saying, 'I am the Messiah'; and many will be misled by them. The time is coming when you will hear the noise of battle near at hand and the noise of battles far away; see that you are not alarmed. Such things are bound to happen; but the end is still to come. For nation will make war upon nation, kingdom upon kingdom: there will be famines and earthquakes in many places. With all these things the birth pangs of the new age begin.'"

"You will be handed over for punishment and execution: and men of all nations will hate you for your allegiance to me. Many will fall from their faith; they will betray one another and hate one another. Many false prophets will arise, and will mislead many; and as lawlessness spreads, men's love for one another will grow cold. But the man who holds out to the end will be saved. And this gospel of the Kingdom will be proclaimed throughout the earth as a testimony to all nations; and then the end will come."

"It will be a time of great distress; there has never been such a time from the beginning of the world until now, and will never be again. If that time of troubles were not cut short, no living thing could survive; but for the sake of God's chosen it will be cut short."

"As soon as the distress of those days has passed, the sun will be darkened, the moon will not give her light, the stars will fall from the sky. The celestial powers will be shaken. Then will appear in heaven the sign that heralds the Son of Man. All the peoples of the earth will make lamentation, and they will see the Son of Man coming on the clouds of heaven with great power and glory. With a trumpet blast He will send out his angels, and they will gather his chosen from the four winds, from the farthest bounds of heaven on every side. When you see all these things, you may know that the end is near, at the very door. Heaven and earth will pass away; my words will never pass away."

"But about that day and hour no one knows, not even the angels in heaven, not even the Son; only the Father."

2Peter 3:8–13: "And here is one point, my friends, which you must not lose sight of; with the Lord one day is like a thousand years and a thousand years like one day. It is not that the Lord is slow in fulfilling his promise, as some suppose, but that He is very patient with you, because it is not his will for any to be lost, but for all to come to repentance."

"But the Day of the Lord will come; it will come, unexpected as a thief. On that day the heavens will disappear with a great rushing sound, the elements will disintegrate in flames, and the earth and all that is in it will be burnt up."

"Since the whole universe is to break up in this way, think what sort of people you ought to be, what devout and dedicated lives you should live! Look eagerly for the coming of the Day of God and work to hasten it on; that day will set the heavens ablaze until they fall apart, and will melt the elements in flames. But we have his promise, and look forward to a new heavens and a new earth, the home of justice."

These amazing statements were made about 2000 years before we knew any thing about thermonuclear reactions. However God knew about these reactions and was giving us warning well in advance of these final events.

There are some people who still believe that the universe will go on forever. An article from the Associated Press, dated Washington DC. May 26, 1999 states, "The expanding universe supports the Big Bang theory, the idea that the universe began when all matter was compressed into a single point that then exploded. The theory states that the universe has been expanding ever since."

"Recent studies also have shown that the universe probably will expand forever. Some researchers also believe that the expansion is speeding up due to a force that accelerates matter. Existence of this force is still controversial."

If the doubters of this force are ever unfortunate enough to be in close proximity to a thermonuclear explosion, they will very suddenly become believers, but they will just as suddenly become dead believers. If the shock wave doesn't kill them, the heat will.

A previous article, in 1998, from the same Associated Press office in Washington DC had this to say, "By measuring the motion of two hundred stars in the Milky Way, researchers at the Max Planck Institute for Extraterrestrial Physics, in Germany, found that stars nearest the Galaxy's center move the fastest, some speeding along at more than 600 miles a second." They are being drawn into a massive black hole whose gravitational attraction is 2.6 times more powerful than our sun.

"This is the strongest case we have yet for a super massive black hole at the center of the Milky Way," astronomer Andreas Eckart said at a news conference of the American Astronomical Society. The concept that a black hole, "Sucking in the stars", exists in the center of the Milky Way has long been controversial and many astronomers rejected some earlier evidence. Is the Universe expanding or collapsing or doing both at the same time? Only God the Father knows, and He has told us to watch the signs of the times and be ready. Even if we knew exactly when the hydrogen would be completely used up, the existence of a super massive black hole could cause the collapse of our galaxy before the fusion reaction stops, so the warning is even more important. **BE READY!**

CHAPTER 18

Epilogue

The Fate of the Believer in Random Chance

Ecclesiastes: 1:1&2: The words of the Teacher, the son of David, King in Jerusalem. Emptiness, emptiness, says the Teacher, emptiness, all is empty.

Ecclesiastes: 12: Remember your Creator in the days of your youth, before the time of trouble comes and the years draw near when you will say, "I see no purpose in them." Remember Him before the sun and the light of day give place to darkness, before the moon and the stars grow dim, and the clouds return with the rain — when the guardians of the house tremble, and the strong men stoop, when the women grinding the meal cease work because they are few, and those who look through the windows look no longer, when the street-doors are shut, when the noise of the mill is low, when the chirping of the sparrow grows faint and the song birds fall silent; when men are afraid of a steep place and the street is full of terrors, when the blossom whitens on the almond-tree and the locust's paunch is swollen and caper-buds have no more zest. For man goes to his everlasting home, and the mourners go about the streets. Remember Him before the silver cord is snapped and the golden bowl is broken, before the pitcher is shattered at the spring and the wheel broken at the well, before the dust returns to the earth as it began and the breath returns to God who gave it. Emptiness, emptiness, says the Teacher, all is empty.

So the Teacher, in his wisdom, continued to teach the people what he knew. He turned over many maxims in his mind and sought how best to set them out. He chose his words to give pleasure, but what he wrote was the honest truth. The sayings of the wise are sharp as goads, like nails driven home; they lead the assembled people, for they come from one shepherd. One further warning, my children: the use of books is endless, and much study is wearisome.

This is the end of the matter; you have heard it all. Fear God and obey His commands; there is no more to man than this. For God brings everything we do to judgement, and every secret, whether good or bad.

The Fate of the Believer in Jesus the Christ

Matthew 11:28–30; "Come to me, all whose work is hard, whose load is heavy; and I will give you relief. Bend your necks to my yoke, and learn of me, for I AM gentle and humble-hearted; and your souls will find relief. For my yoke is good to bear, and my load is light."

John: 3:16–18; "God loved the world so much that he gave his only Son, that everyone that has faith in him may not die but have eternal life. It was not to judge the world that God sent his Son into the world, But that through him the world might be saved.

The man who puts his faith in Jesus does not come under judgement; but the unbeliever has already been judged in that he has not given his allegiance to God's only Son."

John 11:25; "Jesus said, 'I AM the resurrection and I AM life. If a man has faith in me, even though he die, he shall come to life; and no one who is alive and has faith shall ever die."

CHAPTER 19

Summary

The only valid worldview is that GOD is Final Reality; that which leaves no further chance for action, discussion, or change. He is personal and immortal. He has shaped all matter and energy into its present form by the power of his WORD. His name is I AM. He has expressed himself in three forms, the Father, the Son, and the Holy Spirit. He is the creator of all things and as such his Creation does not limit him. He has revealed himself to us in his creation by leaving a record of how He created the universe (background radiation). He has also left a record of how and when He created life (the fossil record and DNA). He has also provided us with true written history of who He is and what He expects from humans. He can and does visit his Creation and at one period in time He took on himself the form of a man, Jesus the Christ.

The other worldview is that the mind of man is final reality and that all matter and energy has been shaped into its present form by random chance. This belief has led to a variety of myths, false beliefs, outright lies, optical illusions, and other misconceptions that have been proposed by dead humans. God through Jesus has also provided the way for humans to be born over again, so that they can learn the truth about Final Reality.

Science is a method of determining probability. This method requires observation, formation of theories about these observations, and experiments that will either verify or falsify the theories.

The term Big Bang was originally an attempt to ridicule the idea of creation. However with the discovery of background radiation and phase transitions in the development of the Universe, it has become obvious that there really was a beginning, and that the Universe has been highly structured from the very beginning. The creation account in Genesis, although very brief, matches the findings of the phase transitions.

CHAPTER 20

The Challenge

Have you ever heard the expression, a big fish in a small pond? It always brings to mind the star quarterback from a small, country town with not even a traffic light to boast about. In my mind's eye, I see him at the counter of a diner twenty years later, on his third coffee warm-up, recalling the glory days to other has-beens. I guess it is simply a matter of perspective: are you a big part of something small or a small part of something big?

The trouble with the big fish is that he has either over estimated his size or under estimated the size of his pond. In the reflection of my dreams, I stand on a shore of the universe, dwarfed by the sands of time and the infinity of expanses. I'm a mere particle and barely measurable, minute in my miniscularity. But, like all of the other specks, I help add up to something, something bigger and more important than myself.

One of the most amazing things about Jesus is that He knew who He was, but He still submitted himself to the public humiliation of a death sentence for you and me. He bowed to His Father's wish and to the plan of salvation in spite of His own desires and vulnerabilities. God did not have to achieve salvation in this way but did so in order that we might, at the very least, have an example to follow. How much more should we forsake ourselves in pursuit of His perfect plan!

> Isaiah 46:10–13: God says, "I reveal the end from the beginning, from ancient times I reveal what is to be. I say my purpose shall take effect. I will accomplish all that I please . . . Mark this; I have spoken, and I will bring it about, I have a plan to carry out, and carry it out I will. Listen to me, all you stubborn hearts, for whom righteousness is far off: I bring my righteousness near, it is not far off, and my salvation will not be delayed."

Yes, you have a choice in how you live! But just as a stone thrown into still waters produces an enlarging rippling effect, so do your choices produce ripples that effect more than just your immediate surroundings. I challenge you to be a small part of something Big!!! But, do not lose sight of this fact; God does not need you in order to complete His plans. He does, however, want you to play a part in them.

Deborah Oliver

All Quotations from the Bible are from *The New English Bible,* Oxford University Press—Cambridge University Press, 1970.

[1] John W. Oiler. Jr. *The Creation Hypothesis* (Downers Grove. Ill.: InterVarsity Press. 1994), p. 245.

[2] *Time,* Microsoft(R) Encarta(R) 96 Encyclopedia. (c) 1993–1995. Microsoft Corporation. All rights reserved. (c) Funk & Wagnalls Corporation. All rights reserved.

[3] Stephen W. Hawking, *A Brief History Of Time* (New York, N.Y. Bantam Books. 1990),
p. 32.

[4] *Science,* 14 July 1972, volume 177, pages 166–170

[5] Paul Chien, *The Real Issue* (Christian Leadership Ministries Volume 15 Number 3 March / April 1997). *Explosion of Life.* pp.1–4.

[6] Franky Schaeffer V. *Addicted to Mediocrity* (Westchester. Ill.: Crossway Books. 1981), pp. 27–28.

[7] Nancy R. Pearcey and Charles B. Thaxton, *The Soul of Science* (Wheaton, Ill: Crossway Books. 1994), p. 165.

[8] Ibid. pp. 184–185.

[9] Stephen Hawking, *A Brief History Of Time* (New York, NY: Bantam Books. 1988), p. 32.

[10] Nancy R. Pearcey and Charles B. Thaxton, ibid. p. 166

[11] David Appell, *Scientific American,* September 2000, *Unlimited Light.*

[12] James Madison, *The Records of the Federal Convention of 1781,* Max Farrand, ed. (New Haven, Conn: Yale Univ. Press, 1911), Vol. I, pp. 450–452, June 28, 1787.

[13] J. Ramsey Michaels, *Inerrancy and Common Sense* (Grand Rapids, Mich.: Baker Book house, 1980), p.68. Footnote 31.

[14] Simon Greenleaf, *The Testimony of the Evangelists* (New York, James Cockcroft & Company. 1874). p. vii–viii.

[15] Ibid. p. 7.

[16] Ibid. p. 9.

[17] Ibid. p. 23.

[18] Ibid. p. 24.

[19] Ibid. p. 25.

[20] Ibid. p. 28.

[21] Francis A. Schaeffer, *A Christian Manifesto* (Westchester, Ill: Crossway Books. 1981), pp. 17–18.

[22] Ibid. pp. 18–19.

[23] Hong-Yee Chiu, *Big Bang Theory* (Danbury, CT., Grolier Inc. *The Grolier Multimedia Encyclopedia* 1995)

[24] George Smoot and Keay Davidson, *Wrinkles in Time,* (New York, NY. William Morrow and Company, Inc. 1993), pp. 283–285.

[25] Ibid, p. 285.

[26] Ibid. p. 291.

[27] Ibid. p. 290.

[28] Ibid. p. 293.

[29] Ibid. p. 291.

[30] Ibid. p. 296.

[31] Ibid. p. 297

[32] Verne Grant, *The Evolutionary Process*. This book is no longer available from the publisher.

[33] Paul Hamlyn, *Hans Christian Anderson's Fairy Tales* Spring House, Spring Place, London NM5, 1959), pp.63–68.

[34] R.Vallery-Radot, 1920. *The Life of Pasteur.* Translated from the French by Mrs. R. L. Devonshire. Doubleday, New York: p. 109.

[35] Francis Darwin, 1887. *The Life and Letters of Charles Darwin*. New York: D. Appleton, II, p. 202. Letter written 1871.

[36] Charles Darwin, 1872. *The Origin of Species, The Sixth Edition*, by permission of the Rare Book Collection of Bryn Mawr College. The Franklin Library, Franklin Center, Pennsylvania, p. 419

[37] A.I. Oparin, 1924. *Proiskhozhdenie Zhizni*, Izd. Moskovski Rabochii, Moscow. *The Origin of Life*. Translation by S. Morgulis. New York: Macmillan. 1938.

[38] J.B.S. Haldane, 1928. *Rationalist Annual* #148: p. 3–10.

[39] J.D. Bernal, 1949. "The Physical Basis of Life", paper presented before British Physical Society, *The Physical Basis of Life*. London: Routledge, 1951.

[40] H.C. Urey, 1952, *The Planets: Their Origin and Development*. New Haven: Yale University Press.

[41] C.B. Thaxton, W.L. Bradley, and R.L. Olsen, 1984. *The Mystery of Life's Origin*. Philosophical Library, Inc. New York: p. 6.

[42] G.G. Simpson, 1964. *Science*, #143, p. 769.

[43] Anon. 1967. *Nature*, #216, p. 635.

[44] Ilya Prigogine, G. Nicolis, and A. Babloyantz, Nov. 1972. *Physics Today*, p. 23–31.

[45] S.L. Miller, 1974, *The Heritage of Copernicus*, ed. Jerzy Neyman. Cambridge. Mass.: The MIT Press, p. 228.

[46] H.P. Klein, July 30, 1976. *The New York Times*.

[47] C.B. Thaxton, W.L. Bradley, and R.L. Olsen, 1984. Ibid. p. 182.

[48] Ibid. p. 185.

[49] Michael J. Behe, *Mere Creation* (Downers Grove, Ill.: InterVarsity Press. 1998. p.183.

[50] Lee M. Spetner, *Not By Chance* (Brooklyn, New York: Judaica Press, Inc. 1998 p. 214.

Suggested Reading

A Christian Manifesto by Francis A. Schaeffer

Wrinkles in Time by George Smoot and Keay Davidson

Not By Chance by Lee M. Spetner

The Mystery of Life's Origin by C.B. Thaxton, W.L. Bradley, and R.L. Olsen

The Creation Hypothesis edited by J.P. Moreland

The Testimony of the Evangelists by Simon Greenleaf

Mere Creation edited by William A. Dembski

The Soul Of Science by Nancy R. Pearcey and Charles B. Thaxton

Rare Earth by Peter D. Ward and Donald Brownlee

The Effects of Nuclear Weapons, AFP 136–1–3, Editor, Samuel Gladstone, published by the United States Atomic Energy Commission.

To order additional copies of

God, Science, and the Big BANG

Have your credit card ready and call:

1-877-421-READ (7323)

or please visit our web site at
www.pleasantword.com

Also available at: www.amazon.com